International Conference on Applied Mathematics, Computational Science and Systems Engineering (AMCSE 2017)

ITM Web of Conferences Volume 16 (2018)

Athens, Greece
6 - 8 October 2017

Editor:

Nikos Bardis

ISBN: 978-1-5108-5576-2

Printed from e-media with permission by:

Curran Associates, Inc.
57 Morehouse Lane
Red Hook, NY 12571

proceedings.com

Some format issues inherent in the e-media version may also appear in this print version.

Printed by Curran Associates, Inc. (2018)

For additional information, please contact EDP Sciences – Web of Conferences at the address below.

EDP Sciences – Web of Conferences
17, Avenue du Hoggar
Parc d'Activité de Courtabœuf
BP 112
F-91944 Les Ulis Cedex A
France

Phone: +33 (0) 1 69 18 75 75
Fax: +33 (0) 1 69 28 84 91

contact-edps@webofconferences.org

Additional copies of this publication are available from:

Curran Associates, Inc.
57 Morehouse Lane
Red Hook, NY 12571 USA
Phone: 845-758-0400
Fax: 845-758-2633
Email: curran@proceedings.com
Web: www.proceedings.com

TABLE OF CONTENTS

Statement of Peer review

In submitting conference proceedings to ITM Web of Conferences, I certify to the Publisher that I adhere to the **Policy on Publishing Integrity** of the journal in order to safeguard good scientific practice in publishing.

1. All articles have been subjected to peer review administered by the proceedings editors.
2. Reviews have been conducted by expert referees, who have been requested to provide unbiased and constructive comments aimed, whenever possible, at improving the work.
3. Proceedings editors have taken all reasonable steps to ensure the quality of the materials they publish and their decision to accept or reject a paper for publication has been based only on the merits of the work and the relevance to the journal.

Title, date and place of the conference

The 2017 International Conference Applied Mathematics, Computational Science and Systems Engineering
AMCSE 2017
Athens, Greece, October 6-8, 2017

Proceedings editor(s):

Prof. Nikos Bardis, Hellenic Military Academy, Greece

Date and editor's signature

December 7, 2017

https://doi.org/10.1051/itmconf/20181601001

Speed control of switched reluctance motor using genetic algorithm and ant colony based on optimizing PID controller

Hassan El-Sayed Ahmed Ibrahim[1], *Mohamed* Said Sayed Ahmed[2], and *Khaled* Mohamed Awad[3,*]

[1]Electrical Engineering and Computer Control Department at Arab Academy for Science, Technology and Maritime Transport, Cairo, Egypt
[2]Electrical Engineering and Control Department at the Military Academy, Cairo, Egypt
[3]Instrumentation and control department at BAPETCO/SHELL J.V Petroleum Company, Cairo, Egypt

Abstract. Proportional-Integral-Derivative control is the most used kind of control which provides the simplest and most effective solution to different kinds of control engineering applications. But until now PID controller is poorly tuned in real life and online applications. While most of PID tuning is done manually. Switched reluctance motor (SRM) has highly nonlinear characteristics since the developed/produced torque of the motor has a nonlinear function on both phase current and rotor position. These nonlinearities of the SRM drives make the conventional PID (proportional + integral + Derivative) controller a poor choice for application where high dynamic performance is desired under all motor operating conditions. research paper comes up with two artificial and hybrid techniques involving Genetic Algorithm (GA) and Ant Colony Optimization (ACO). Those techniques where used to tune the PID parameters for the switched reluctance motor (SRM) and its performance were compared with the conventional method of "Ziegler Nichols. The results obtained reflects that, the use of those algorithms based controller improves the performance of the whole process in terms of a fast set point tracking and regulatory changes and also provides an optimum stability for the system itself with a minimum overshoot on the output signal.

1 Introduction

In this research, we show that the control of the switched reluctance motor (SRM) is apparently an easy task but in real world it is difficult to be achieved when using a simple PID controller due to its nonlinear characteristics [1, 15]. So according to that, we use artificial intelligent techniques like Genetic Algorithm and Ant-Colony optimization to be able to achieve a robust steady speed performance under different load circumstances.

Also it has some good features, like high power density, high efficiency and no rotor windings. This is one of its major advantages. However, the SRM motor suffers from torque ripples and acoustic noises that prevent its use in high performance drives [14].

A Matlab-Simulink model for the speed control of a 3-phase switched reluctance motor carried out using different controllers. The Simulink models designed here are the traditional PID, Genetic-PID and Ant Colony-PID controllers which are designed separately and their performance results were compared together. The speed controller applied here is based on the traditional PID Controller and the other one is an artificial intelligent based controller like Genetic algorithm & Ant Colony Controllers. A comprehensive review/research has been done for SRM machine modelling, simulation, analysis and control. The conventional controller design (*i.e.* PID) is based on the mathematical model of the plant, which may often be unknown, ill-defined, nonlinear and complex. The PID controller has fixed gains and is certainly the most commonly used control algorithm nowadays. The main reason for using it until now is its relatively simple structure which can be easily understood and implemented in real world. The PID controller has traditionally been tuned empirically e.g. by the method described in Ziegler and Nichols. The controller gains once tuned for a given operating point for a specific plant or a system are only suitable for limited operating point changes and this is one of its disadvantages. Therefore, the use of the conventional PID controller does not meet the requirements of a robust & steady speed performance. Various control strategies has been proposed for the speed control of the SRM, such as feedback linearization control, fuzzy logic control and neuro-fuzzy control. While GA and ACO techniques those are based on the behavioral pattern of a living being and are a new artificial methods introduced strongly to control the SRM motor. The objective of this research paper is to show that a 3-phase SRM motor using the conventional PID controller, its values can be obtained and optimized using the artificial techniques of

* Corresponding author: eng.kha_Awad@hotmail.com

GA and ACO, which gives better results with sudden load changes and system stability.

2 Switched Reluctance Motor

The switched reluctance motor has a passive rotor with a simple construction. While the solution of its mathematical model is relatively difficult due to its non-linear behavior. The SRM is characterized by its geometrical Layout, the characteristic of the magnetic material and electrical parameters Switched reluctance motor (SRM) drives are of simple construction compared to the induction or synchronous motors. The structure of the motor is very simple that uses salient poles on both stator and rotor but only stator carries winding. The windings on the stator are of simple form. So this apparent simplicity of its design is not very accurate because its construction is simple but controlling the motor itself is difficult. SRM drives has some advantages as it has high efficiency, maximum operating speed, good performance of the motor in terms of torque/inertia ratio, according to that it can be used in variable speed applications with high speed. The performances of switched reluctance motor strongly depend on its applied control. Figure 1 shows the major parts of a switched reluctance motor drive.
There is three main parts identified as follows: the motor itself, the power electronic converter and the controller used. SRM construction consists of the following: There is a single converter per phase; a battery supplies the dc power to the converter. Its operation is simple: When the excitation is given to the stator windings, a force is created by rotor's magnetic reluctance that tries to align the rotor pole with its adjacent stator pole (Fully aligned) in order to maximize the inductance & minimize reluctance of the excited coils. In order to preserve the correct sequence of rotation, the windings of stator pole switches in a sequential order with the help of electronic control system so that the magnetic field of rotor pole was led by the stator pole, pulling towards it production of torque is independent of the direction of current which in-turn reduce the number of switches per phase.

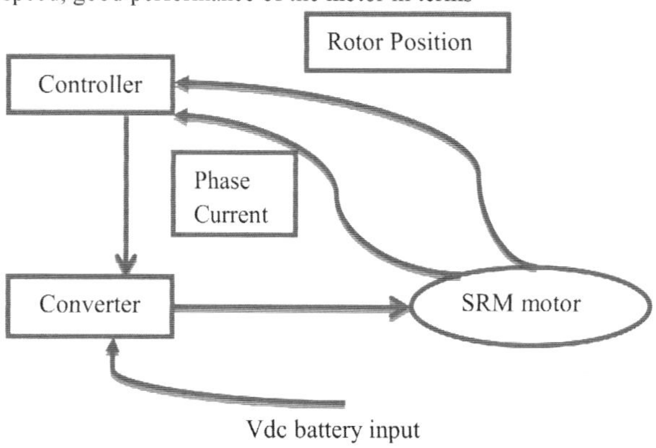

Fig.1. Switched Reluctance Drives System Components

3 SRM Block Diagram

The position of rotor is sensed by the rotor position sensor and it provides its corresponding output to the error detector. Error detector compares reference speed given in RPM and the actual speed feedback from the motor itself to generate an error signal which is given to the controller block. The controller (PID or GA-PID or ACO-PID) gives control signal to the converter according to the error signal value. The speed of the motor is controlled by the converter through proper excitation of their corresponding windings.

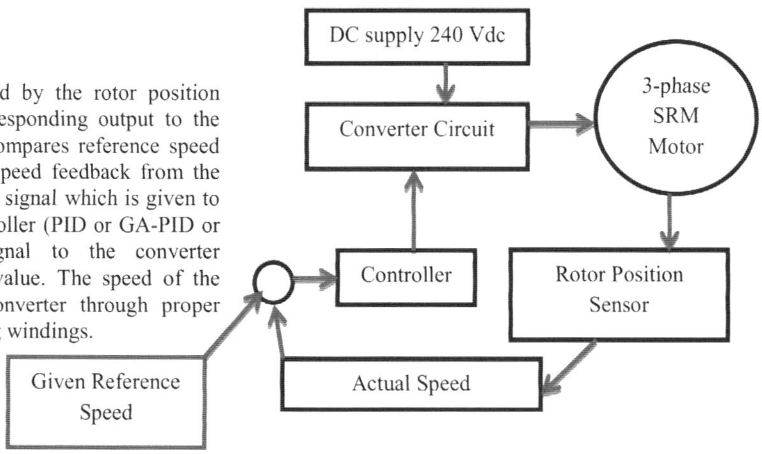

Fig.2. Block Diagram of SRM Speed Control

 https://doi.org/10.1051/itmconf/20181601001

4 Genetic Algorithms

Genetic Algorithms were invented to try to mimic some of the processes observed with living being [3-5]. The idea with GA is to use the power of natural living ability for evolution to solve hard and very complicated Engineering optimization problems [13]. GA's process starts with no knowledge at all for the correct and accurate solution and depends entirely on responses it receives from the surrounding environment and its internal evolution operators such as reproduction, crossover and mutation to get to the best available solution.

The application of these three basic operations allows the creation of new and better individuals, which may be better than their old parents. This algorithm is repeated for many generations and at the end we stops when we reaches for individuals that represent the optimum solution to the problem or we stop at certain number of iterations we initially stated.GA starts with an initial and random chromosome & checks its fitness value. The fittest chromosome are taken as parents for further reproduction also to be crossed over & mutated again. The offspring or their children are checked for the fitness value & depending on it. We can decide either to be taken again or to be neglected from the population [7, 16, 17]. Genetics father is "John Holland" who invented this technique in the early 1970's [6]. GA's is an adaptive artificial search technique based on the evolutionary ideas of natural selection [7]. In order to use the "GA's" there are two important aspects to follow: chromosome coding and defining the evaluation criteria. Although it appears as a randomized process, GA's is not a completely random process. Instead it develops some historical information to direct the search into the region of better performance within the search space it has. The Basic techniques of the GA's are designed to simulate processes with natural living being which is necessary for evolution as found on the block diagram on Figure 3. Especially those who follows the principles of Charles Darwin "survival of the fittest" Since in nature, competition among individual's results in the fittest individuals dominating over the weaker ones and this is observed and proved its existence in history. GA's can give robust & adaptive response for a system with nonlinearity, parameter variation and load disturbance effect [3-5].

GA's is better than traditional techniques and more robust to sudden load changes and is a very adaptive technique. Unlike old AI systems, they do not break easily if the input changes or noise applied suddenly. Also in searching a large state-space, GA's may offer significant benefits over the typical search of optimization artificial techniques.

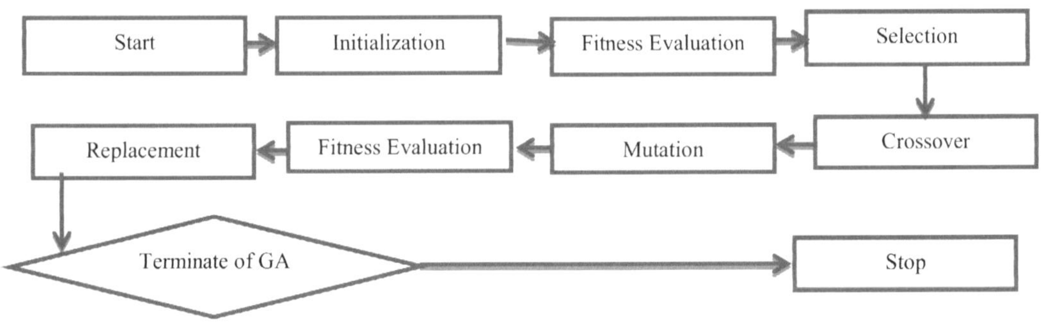

Fig.3. Block Diagram for GA Sequence

5 Ant Colony Optimization

Marco Dorigo and some of his colleagues invented the first ACO concept in the early 90's [8, 9, 10]. The development of these algorithms was inspired by the observation of ant colonies in nature. Ants are social insects with an ability to organize their food obtaining process very accurately. They live in colonies (groups) and their behavior is governed by the goal of colony survival rather than being focused on the survival of individuals. Ant algorithms are adaptive and robust to any complicated engineering problem. The behavior that provided the inspiration for ACO is the ants "foraging behavior" and in particular, how ants can find the shortest path between food sources and their nest (Home). The main advantages of artificial ants are taken from their natural model. Which are (1) artificial ants exist in colonies of cooperating individuals; (2) they communicate indirectly by depositing pheromone on their way to any expected food sources [11, 12] (3) they use a sequence of local moves to find the shortest path from a starting position which is completely random, to a destination point they apply a stochastic decision policy using local information only to find the best solution.

When searching for food supplies (Sources), ants initially searches the surrounding area or explores it which is near its nest. This is in completely random directions. While moving it leaves a small amount of pheromones behind it on the ground. This is in order for the other ants to smell it during this trip of finding food sources. While other ants search for food it chooses its way in a probability which is according to the amount of pheromones it finds. As soon as an ant finds a food source, it tries to know the quantity and the quality of these sources of food and carries some of it back to its home base. During the return trip the quantity of pheromone it was left over during the trip is evaluated again to try to know the shortest route to its nest. When it reaches its nest other ants will go on the same route to the food source which contains actual food as illustrated on figure 4.

It has been shown that, this kind of communications is an indirect one which is built in every ant to get to its nest as soon as possible and in an organized manner also it is called "stigmergy effect" which enables them to find the shortest path between their home base and expected food source.

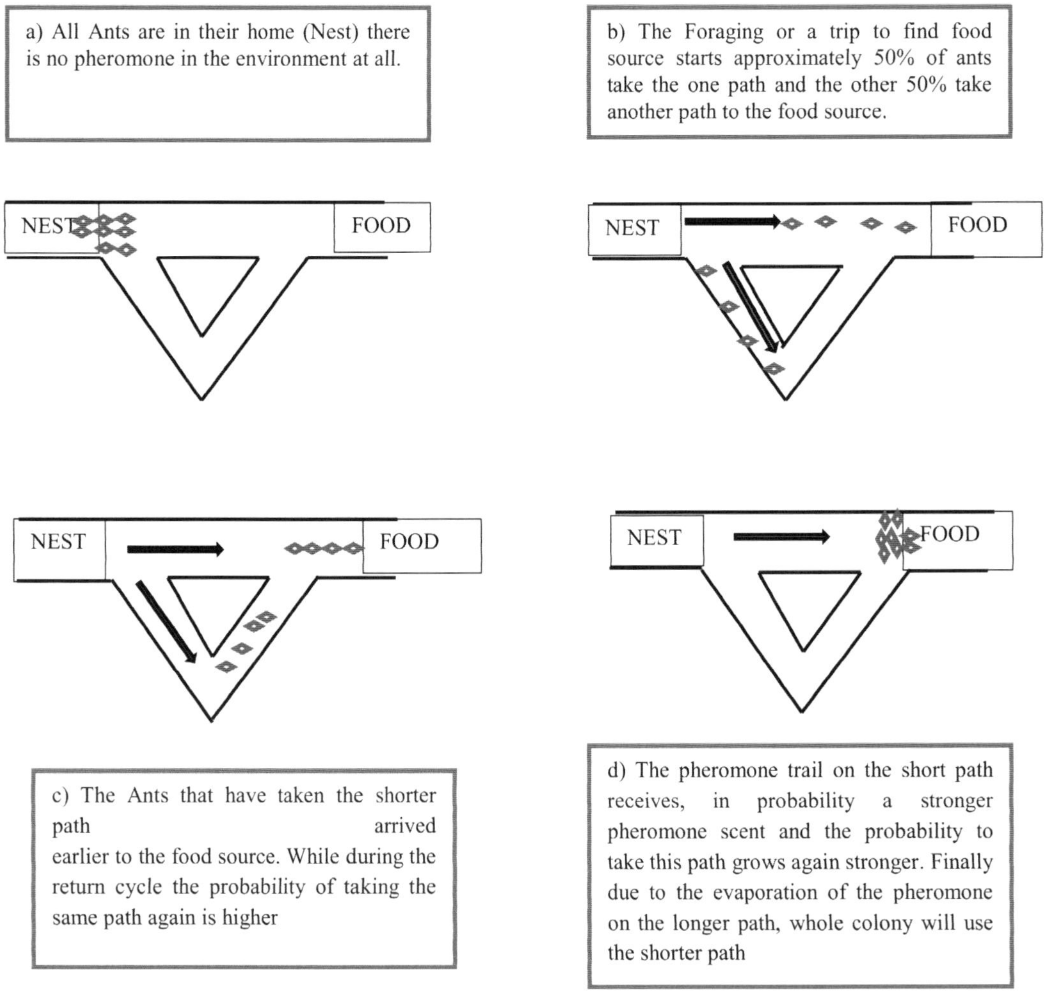

Fig. 4. Experimental setting that demonstrate the Shortest Path Finding Capability of Ant Colonies, between the ants nest and the only food source exist with two paths of different length

6 Experimental Setup of the SRM Drive Using Matlab-Simulink

The experimental setup is shown on Figures 5, 6. This shows the PID controller optimized by GA's also the same PID controller optimized by ACO technique. We start the simulation by using a 3-phase 60KW SRM motor which is used on this research paper with motor features as found on table 1.

Also it is the same built-in Matlab motor model. Tables 2 and 3 shows the GA's and ACO controller's parameters used after several trials. the response of Kp, Ki and Kd parameters are summarized on table 4 which shows that using the proportional controller "Kp" will have the effect of reducing rise time of the output signal and will reduce the steady state error while the integral controller "Ki" will have the effect of eliminating the steady state error have the effect of eliminating the steady state error produced by the "Kp" parameter but will have the worst transient response on any system.

While the derivative controller "Kd" will have the effect of increasing the stability of the system, reducing the overall overshoot and improves the transient response of the system. Several simulations were carried out for controlling motor's speed with a different load torque disturbances to explore the effectiveness of the proposed controller's robustness in comparison with the conventional old PID controller.

The results were obtained for cases such as step variable (Input) speed of ranges starts from "1000-1500 RPM" which is our maximum RPM for such motor and under variable load torque "Nm".

Table 1

Matlab built-in Motor Features

Item	Features Description	Value
1	Rotor Pole number "Nr"	6
2	Stator Pole Number "Ns"	8
3	Turn on angle	40 degree
4	Turn off angle	75 degree
5	Number of Phase	3
6	Power in (KW)	60
7	Maximum (Reference) speed (RPM)	1500
8	Stator Resistance (Ohm)	0.05
9	Inertia (Kg.m.m)	0.05
10	Friction (N.m.s)	0.02
11	Source Voltage (Vdc)	240

Table 2

Genetic Algorithm Parameters

Item	Description	values
1	Proportional gain limits Kp	Ranges [0-100]
2	Integral gain limits Ki	Ranges [0-100]
3	Derivative gain limits Kd	Ranges [0-100]
4	Population type	Double vector
5	Population size	100
6	Creation Function	Uniform
7	Mutation rate	0.1
8	Mutation Function	Uniform
9	Selection method "Function"	Uniform
10	Crossover type	Arithmetic

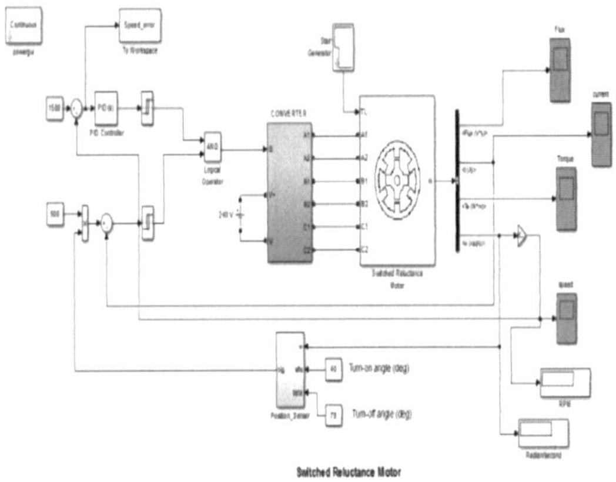

Fig.5. SRM motor speed control with GA-PID Controller.

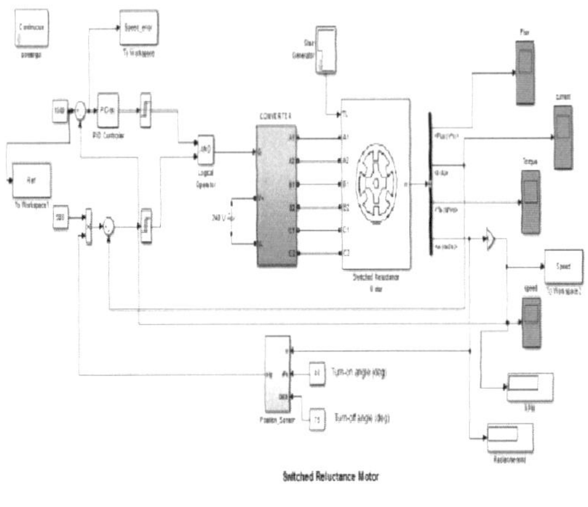

Fig.6. SRM motor speed control with ACO-PID controller

Table 3

Ant Colony Parameters

Item	Description	values
1	Number of Iterations	100
2	Number of Ants	300
3	Alpha	0.8
4	Beta	0.2
5	Evaporation Rate	0.7
6	Number of Parameters	3
7	Lower& upper limits	[000] & [111]

Table 4

Response of K_p, K_i and K_d

Item	K_p	K_i
Rise time	Decrease	Small change
Overshoot	Increase	Decrease
Settling time	Small change	Increase
Steady state error	Decrease	Eliminate
Item	K_p	K_i

7 Simulation Results

To explore the effectiveness of the proposed techniques, a Matlab-Simulink computer simulation has been carried out for the different proposed controllers. The conventional PID, GA-PID and ACO-PID controllers system [*i.e.* the maximum torque obtained, rotor's position, reference and actual speed values, speed error signal and change in the error signal, input values of torque, produced flux current] can be observed.

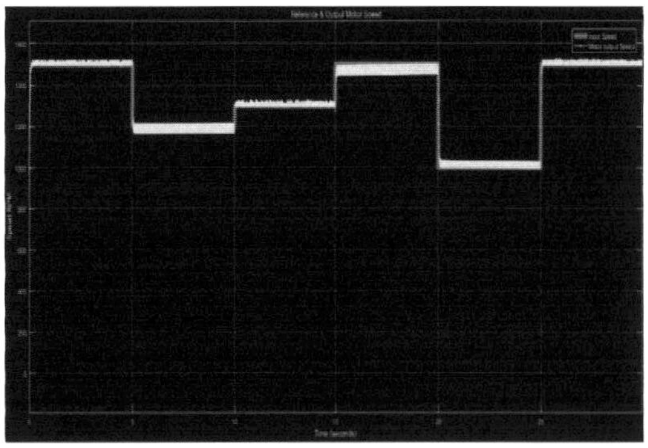

Fig.7. shows the variable step input (Speed) and the motor actual output speed after using Conventional PID controller.

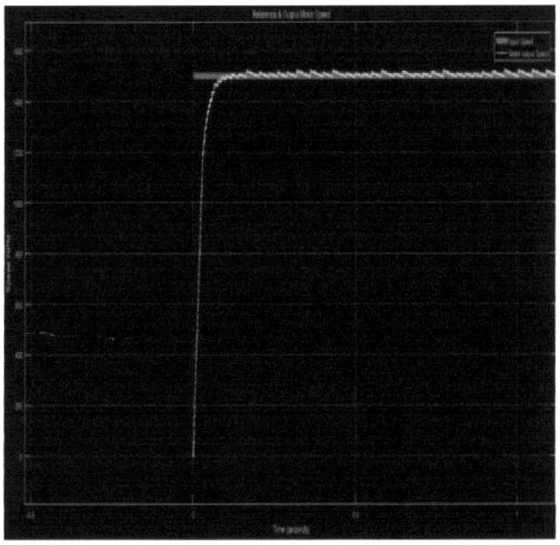

Fig.8. shows the delay on the output (Zoomed in) of the motor with reference step input when using conventional PID controller.

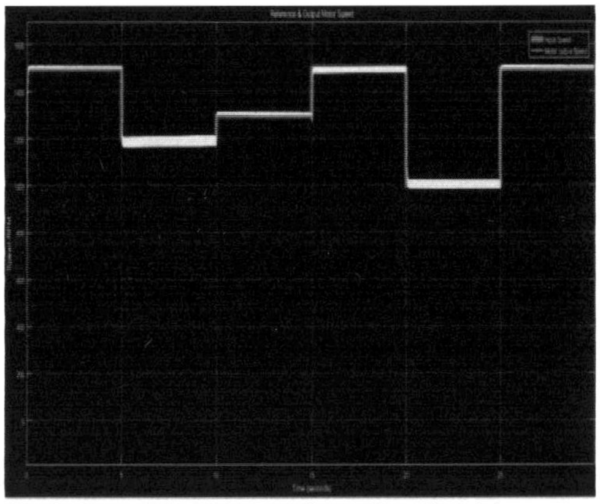

Fig.9. shows the variable step input (Speed) and the motor actual output speed after using the new artificial Genetic algorithm Technique.

https://doi.org/10.1051/itmconf/20181601001

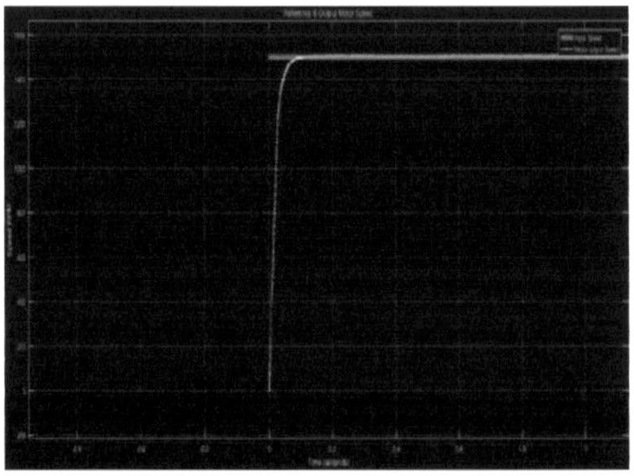

Fig.10. Shows the delay on the output (Zoomed in) of the motor with reference step input when using the Genetic algorithm

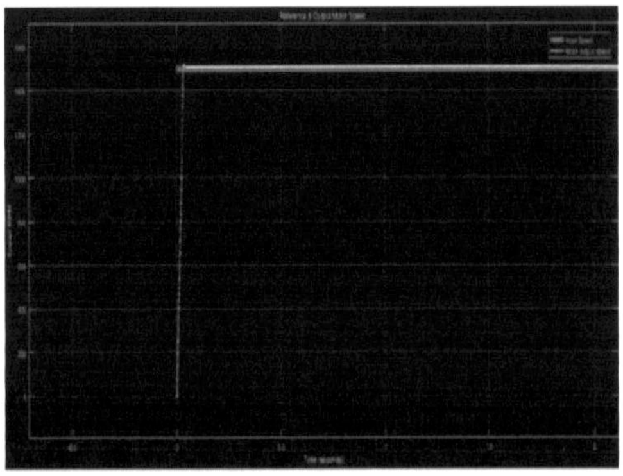

Fig.12. Shows the delay on the output of the motor with reference step input when using the Ant Colony technique

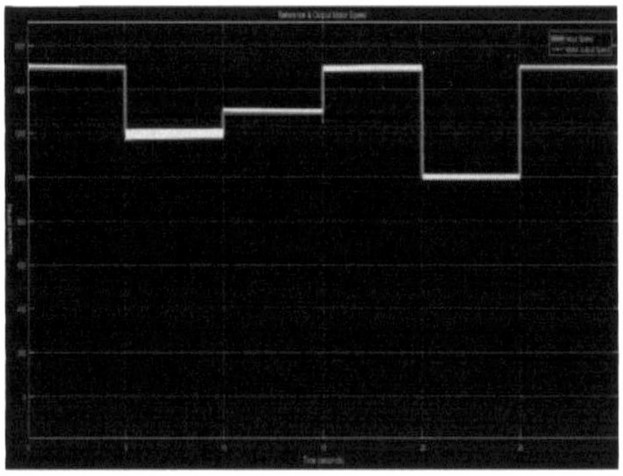

Fig.11. Shows the variable step input (Speed) and the motor actual output speed after using the Ant Colony technique.

8 Future Works

The ACO and GA can be modified for future enhancement such that a new and updated research could and can be focused on each technique disadvantages in order to produce better solution by improving the effectiveness of getting the best solution and reducing its limitations. Research work has been carried out to get the maximum and optimal values for PID controller parameters automatically using GA and ACO artificial optimization techniques as found on figures 9, 10, 11 and 12. Simulation results demonstrate that the tuning methods proposed and used as found on both figures 10 and 12 have a better control performance compared with the conventional ones as found on figure 8 also they have a better steady state response, rise time, settling time, overshoot and a robust control with variable loads as found on table 5.

Table 5

Comparison between the conventional control method (PID) and Heuristic Intelligent artificial methods (GA & ACO)

Tuning Method	PID Parameters			Dynamic Performance Specifications		
ZN	Kp Proportional gain	Ki Integral gain	Kd Derivative gain	Tr Rise time	Ts Settling time	Mp Peak overshoot
	0.95	0.05	0.01	0.070 sec	0.10 sec	1.53 %
GA	282.9	26	2.346	0.060 Sec	0.080 Sec	0.53%
ACO	268.8579	65.86439	0.01	0.036 Sec	0.037 sec	1.2%

REFERENCES

1. Krishnan, R., Switched Reluctance Motor Drives, CRC Press, Boca Raton, Florida, (2001).
2. Ziegler, J. G., and Nichols, N. B., "Optimum settings for automatic controllers," Trans. ASME, Vol. 64, pp. 759–768, (1942).
3. Fleming, P. J., and Purshouse, R. C., "Evolutionary algorithms in control system engineering: A survey," Control Eng. Practice, Vol. 10, pp. 1223–1241,(2002).
4. Krohling, R. A., and Rey, J. P., "Design of optimal disturbance rejection PID controllers using genetic algorithms," IEEE Trans. Evol. Comput, Vol. 5, pp. 78–82, (February 2001).
5. Karr, C., and Gentry, E., "Fuzzy control of ph. using genetic algorithms," IEEE Trans. Fuzzy Syst., Vol.1, No. 1, pp. 46–53, (1993).
6. Goldberg, D. E., Genetic Algorithms in Search, Optimization and Machine Learning, Addison- Wesley Publishing Company, Boston, Mass., (1989).
7. Zhu Supeng & Fu Wenxing Yang Jun Luo Jianjun "Applying Genetic Algorithm to Optimization Parameters of Missile Control System" 978-0-7695-3745-0/092009 DOI 10.1109/HIS.2009.297 IEEE (2009) Ninth International Conference on Hybrid Intelligent Systems.
8. Dorigo M, Stützle T. Ant Colony optimization. Cambridge, MA: MIT Press; (2004).
9. Dorigo M, Optimization, learning and natural algorithms. PhD thesis, Dipartimento di Elettronica, Politecnico di Milano, Italy, (1992) [in Italian].
10. Dorigo M, Blum C. Ant colony optimization theory: A survey. Theoret Comput Sci (2005); 344(2–3):243–78.
11. Deneubourg J-L, Aron S, Goss S, Pasteels J-M. The self-organizing exploratory pattern of the Argentine ant. J Insect Behavior (1990); 3:159–68.
12. Grassé P-P. La reconstruction du nid et les coordinations inter-individuelles chez bellicositermes natalensis et cubitermes sp. La théorie de la stigmergie: Essai d'interprétation des termites constructeurs. Insectes Sociaux (1959):6 41–81.
13. Ian Griffin, Jennifer Bruton "On-Line PID controller tuning using genetic algorithm". Available at: www. eeng.dcu.ie/~brutonj/Reports/IGriffin_MEng_03.pdf.
14. JIn-Woo Ahn (2011). Switched Reluctance Motor, Torque Control, Prof. Moulay Tahar Lamchich (Ed.), ISBN: 978-953-307-428-3, InTech, Available from: http://www.intechopen.com/books/torque-control/switchedreluctance- Motor
15. V. Vasan Prabhu, K. S. Mahesh, C. Renuka,"Simulation of Switched Reluctance Machine for Linear and Nonlinear Model", International Conference on Computer, Communication and Electrical Technology-ICCCET (2011), 18th & 19th March, 2011,pp.333-339

Wavelet Co-movement Significance Testing with Respect to Gaussian White Noise Background

Jitka Poměnková[1,*], *Eva* Klejmová[1], and *Tobiáš* Malach[1]

[1] Brno University of Technology, Faculty of Electrical Engineering and Communication, Technická 12, 616 00 Brno, Czech Republic

Abstract. The paper deals with significance testing of time series co-movement measured via wavelet analysis, namely via the wavelet cross-spectra. This technique is very popular for its better time resolution compare to other techniques. Such approach put in evidence the existence of both long-run and short-run co-movement. In order to have better predictive power it is suitable to support and validate obtained results via some testing approach. We investigate the test of wavelet power cross-spectrum with respect to the Gaussian white noise background with the use of the Bessel function. Our experiment is performed on real data, i.e. seasonally adjusted quarterly data of gross domestic product of the United Kingdom, Korea and G7 countries. To validate the test results we perform Monte Carlo simulation. We describe the advantages and disadvantages of both approaches and formulate recommendations for its using.

1 Introduction

Measuring the mutual dependence and co-movement between two inputs (time series or signals) is at the forefront of the interest of many scientific areas such as economy, sociology, biology, geoscience, climatology, engineering, etc. The selection of appropriate analytical instrument comes out from the application area which predefines analytical requirements. In the last decade it became more common an implementation of methods derived from the technical or nature sciences into the social sciences. Therefore, the need of an examination of the possibilities, of an adaptation and the utilization of such a method, the ways of its application and the correct interpretation of the obtained results arise.

The wide range of the methodology implementation can be found in the economics in analysis of globalisation between economies, emerging and financial markets, regional similarities, analysis of relation of economic, sociologic or demographics indicators or analysis of factors influencing dissemination of innovation, economic growth and many others. For the long time these analyses were predominantly done in the time domain. Empirical work mostly used and still uses regression and correlation methods [1, 2, 3], cointegration [4], codependence, causality monitoring, structural and vector models [5], and so on. Thus, examination of frequency characteristics was rather suppressed or marginal. In some cases the scientists in economy consider the dynamic behaviour as the motion of analysed inputs in terms of its leads and lags [6, 7]. Another option is the frequency characteristics, such as dynamic correlation, which complement the knowledge from time domain in a suitable way and specify and validate the results from the time domain analysis [8-11].

Interconnection of analyses in both domains provides time-frequency (TF) techniques. Commonly used TF methods are Fourier Transform (FT), modifications of periodogram [12], TF varying autoregressive process (TFAR) [13], and continuous wavelet transform (CWT) [14, 15]. The last one becomes very popular, especially in the last decade, for its good time resolution. We can examine in TF domain not only transform of inputs, but also co-movement of inputs. It can be done via instruments such as cross spectrum (WCS), co-spectrum or cohesion. These approaches propose more precise evaluation of co-movement to capture temporal character of co-movement [16-18] then simple time domain techniques.

If we focus on exploring economic phenomena, CWT is at the forefront of economic interest [19, 20]. Application of CWT helps with modelling and exploration of the joint movement of economic indicators. The advantages of CWT can be summarized in the following points: i) applicability on stationary and non-stationary time series; ii) possibility of time and frequency analysis in one, i.e. spectrogram or co-spectrum analysis; iii) flexible choice of mother wavelet with respect to character of the inputs, iv) time resolution is adjusted to the scale, v) compared to other TF techniques, there is no need to optimize the parameters and thus limiting the quality of the resulting TF estimate. There is only discussion about wavelet and scale selection. As the disadvantage we can mention a more complex way of testing and identifying significant areas which is used for interpretation of the results.

* Corresponding author: pomenkaj@feec.vutbr.cz

Together with implementation TF method into various disciplines arises the necessity of validation of the results. The basic works, giving instrument for significance testing, gives Torrence and Compo [21] followed by Ge [22]. Both works propose algorithm for testing wavelet power spectrum, cross-spectrum and linear coherence. Ge [22, 23] proved that the sampling period factor should be incorporated into testing to achieve numerically stable results. In all cases, authors' uses testing statistic with devoted distribution corrected with respect to the background spectrum distribution and sampling period. An alternative approach, presented in [15] or [14], uses Monte Carlo (MC) simulation for identification of critical values.

Based on the real data, we investigate the behaviour of testing procedure for the estimated wavelet cross spectrum with respect to the Gaussian white noise (GWN) background. We use both the statistical significance testing according Ge [22], and the identification of critical values using MC simulations. We deal with the advantages and disadvantages and provide recommendations for usage of TF co-movement significance testing. We follow up the paper [17] which works with significance testing of wavelet coefficients.

The paper is organized as follows: after the introduction, in the section Methodology, we introduce the theoretical background for co-movement testing. In the next section, Application, we describe data and present and summarize the achieved results. The paper ends with the Conclusion.

2 Methodology

In this section we present theoretical background for testing cross spectrum significance. We follow work of Torrence and Compo [21] and Ge [22]. Firstly, we present definition of wavelet cross spectrum (WCS). Consequently, we describe algorithm for statistical testing according to Ge [22] and according to MC simulations.

2.1 Wavelet cross spectrum (WCS)

An analysis of two time series co-movement can be done via WCS. The WCS between two time series/signals can be define as

$$C_{xy}(a,b) = W^*_x(a,b)\, W_y(a,b), \qquad (1)$$

where $W_x(a,b)$ and $W_y(a,b)$ are the coefficients of wavelet transform corresponding to the time series $x(t)$ and $y(t)$ which are both independent GWN with corresponding variances σ_x^2 and σ_y^2. The symbol * denotes complex conjugation, a and b are scale and time variables [22]. Since the WCS is complex, such in our case, when the complex Morlet wavelet are used, we can define its square absolut value, i.e. wavelet power cross spectrum, as

$$|C_{xy}(a,b)|^2 = |W_x(a,b)|^2 |\, W_y(a,b)|^2, \qquad (2)$$

where $|W_x(a,b)|^2$ and $|W_y(a,b)|^2$ are the squared absolut value of WCS, i.e. wavelet power spectras [21].

2.2 Significance test

Let us have two time series $x(t)$ and $y(t)$, which are both independent GWN with corresponding variances σ_x^2 and σ_y^2, and wavelet spectra $W_x(a,b)$ and $W_y(a,b)$. As presented in [21], the square of the absolut value of normally distributed variable is $\chi^2(\nu)$ distribution with ν=2 degrees of freedom. As presented in [22], Wells et al. [24] obtained the probability distribution of two independent non-central χ^2-distributed random variable. In the special case, when both wavelet power spectras are $\chi^2(2)$ and non-centrality parameters are both zero, i.e. when

$$|W_x(a,b)|^2 / (\delta t\, \sigma_x^2 /2) \sim \chi^2(2), \qquad (3)$$

$$|W_y(a,b)|^2 / (\delta t\, \sigma_y^2 /2) \sim \chi^2(2), \qquad (4)$$

where δt is the sampling period, for their product holds

$$|C_{xy}(a,b)|^2 / (\sigma_x^2\, \sigma_y^2) \sim 0.25\delta t^2\, W_2, \qquad (5)$$

where, based on Wells et al. [24], W_2 denotes probability distribution with probability density function

$$f(z)=0.5\, K_0(z^{1/2}). \qquad (6)$$

Here z is the random variable, and $K_0(z)$ is the modified Bessel function of order zero when the complex wavelets are used [22, 24].

2.3 Monte Carlo (MC) simulations

The idea of MC simulations is based on repeated random sampling to achieve numerical results. It uses statistics of original data for the subsequent replication of original data by generating large number of random realization of a stochastic process. On the basis of law of large numbers we can identify empirical distribution close to probabilistic distribution and identify empirical statistics, such as quantile or percentile. Resulting simulated process is consequently approximated by averaging these sub realizations in some way. We are going to apply MC simulations consisting from following steps [25]:

- Step 1: Definition of the model to be simulated
- Step 2: Generating random values of predefine distribution
- Step 3: Calculation with large number of repetitions
- Step 4: Analysis of results, quantiles identification

3 Application

For the demonstration we selected real economic data rather than their simulation, because such real data contain structural breaks and reactions on economic even. In case of its simulation some of unpredictable factor can be supressed, which is not welcomed.

Our approach consists from several steps. In the first step we estimate WCS and wavelet power cross

https://doi.org/10.1051/itmconf/20181601002

spectrum (eq. (2)). In the second step we identify significant regions according to Ge [22] approach (eq. (5)). In the third step we used 1.000 Monte Carlo (MC) simulation to identify empirical critical values. At the end we compare both results (according to Ge and MC simulation) and formulate recommandations.

3.1. Data

For the demonstration of discussed approach we use seasonally adjusted quarterly data of gross domestic product (GDP), volume index in OECD reference year 2010 (OECD [26]) of the United Kingdom (UK) in 1956/01-2016/03, Korea in 1970/Q2-2016/03 and Group of 7 (G7) in 1961/02-2016/03. All variables are in the first differences of the logarithms (Fig. 1a-c). The motivation for data selection were: i) sufficient data range (it is desirable to have detailed time resolution); ii) in case of UK and G7 we expect co-movement, because UK is a member of G7 and thus support validation of purposed method; iii) in case of Korea we expect lower level of co-movement with G7, because Korea is not the member of G7. Additionally, with respect to the current situation in Great Britain (tendency to withdrawal from the EU) we were interested into analysis before such even having character of structural break which will affect the data.

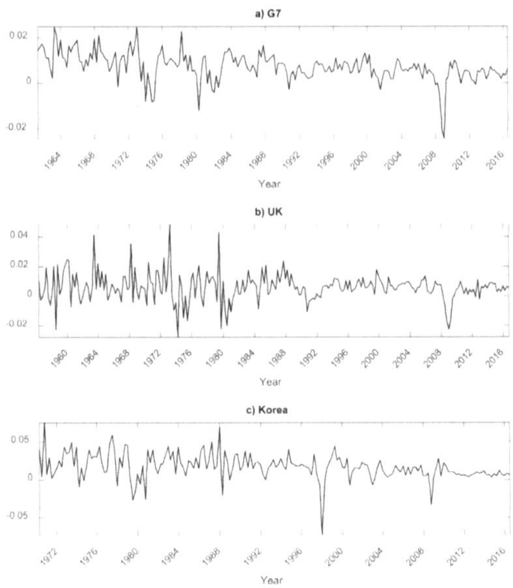

Fig. 1. Time series representation (1a: G7, 1b: UK, 1c: Korea)

3.2 Setting of TF methods and MC

For CWT estimation we used complex Morlet wavelet with central wavelet frequency f_0=1.5 as the mother wavelet. We set scales corresponding to the range from 2 years to 40 years cycles, with 334 individual scales. The power WCS were calculated according to the formula (2). The resulting figures for the data can be seen in Fig. 2a-c.

In case of MC simulations we firstly generate two independent GWN series. Then, we estimate its CWTs and consequently power WCSs with 334 individual scales (corresponding to the data setting). Consequently, we generate 1.000 repetitions with the length of 1.000 points. Not to complicate things for the CWT with cone of influence we selected a time point in middle of WCS spectrogram. For each frequency (scale) from all MC iterations we save the obtained values. After that, we calculated 95% quantile of empirical distribution separately for each vector (scale).

3.3 Results

The results of co-movements are presented in Fig. 2a-c below. All figures have x-axis: the time, y-axis: the frequency measured in the cycles per year and z-axis: the power WCS. The y-axis was re-calculated from frequency to the cycles per year for the better interpretation of an economic inputs. Thus, given the quarterly character of the data, we denote the sampling frequency F_s to be 4 samples per year. The red curve in the figures defines significant area identify via the MC simulations, the yellow dashed curve defines significant area according to Ge [22].

Focusing on the results of co-movement (Fig. 2a-c) with respect to the identification of significant area via Ge [22] we can state following. All figures show existence of co-movement between countries which matches our assumptions. That is, the UK-G7 (Fig. 2a) has higher level of co-movement caused by being UK in G7. The Korea and G7 (Fig. 2b) has less areas with co-movement, but generally both UK-G7 and Korea-G7 has co-movement in the time of oil crisis 1970-1982 and in the time of financial crisis 2001-2010. Such strong co-movement of Korea with G7 is also visible in Fig. 2c. All these significant co-movement (dashed curves in the Fig. 2a-c) were identified for long cycles, i.e. 10 years length cycles and longer. These results were confirmed by MC simulations.

Focusing on same figures (Fig. 2a-c), but on significant areas identified by the MC simulation, we can see an additional areas in all measured co-movements. That is, the additional co-movement between UK-G7 (Fig. 2a) in the time of oil crisis 1968-1982, but for shorter cycles, i.e. for the cycles of the approximately 2-4 years. Such additional co-moved area reflects the faster response in the country economy to the oil crisis, such as small and medium-sized enterprises, loans, investments etc. In the case of the co-movement between Korea-G7 (Fig. 2b) we can see also additional co-movement in the period 1974-1982 but for cycles of the length 4-6 years. Similarly, between Korea-UK (Fig. 2c) there is also the additional co-movement in the period 1972-1980 for cycles of the length 4-6 years. Here, the results again confirm our expectation of direct interconnection of UK and G7 and the weaker interconnection of Korea and G7. In all the cases results confirm the globalisation of the economies and the stronger impact of the oil crisis on the world economy then the financial crisis.

 https://doi.org/10.1051/itmconf/20181601002

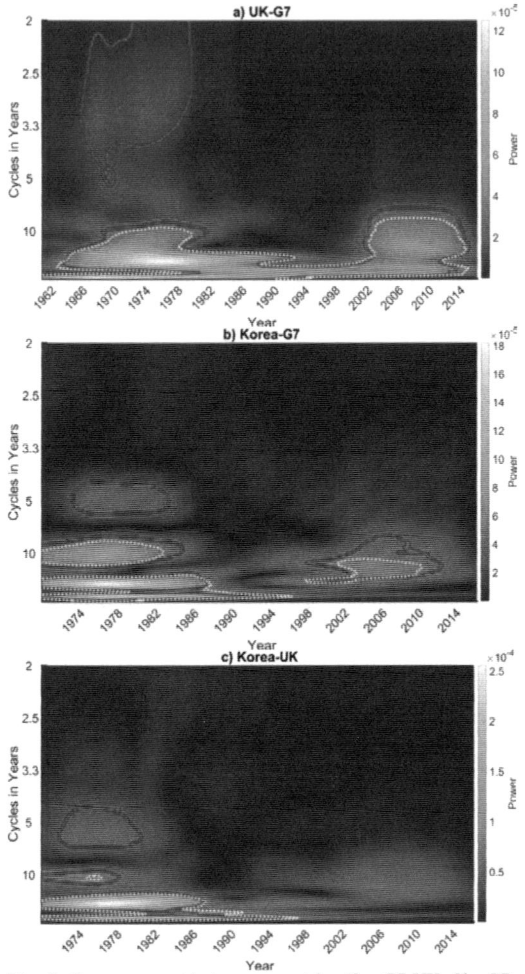

Fig. 2. Co-movement between countries (2a: G7-UK, 2b: G7-Korea, 2c: Korea-UK)

The comparison and the verification of the accuracy and suitability of Ge approach via the MC simulations is presented on Fig. 4. For the illustration of the significance level we firstly preselect the three time slices of the WCS (for example of UK-G7 (Fig. 3)). Selected time slices are denoted by the numbers 1-3 and corresponds to the time 1976, 1998 and 2006. The figure (Fig. 4) demonstrates power WCS curves and the significance level identified by Ge (dashed-dotted line) and by the MC simulations (stars). We can see that the MC significance level is not a constant, resp. it is decreasing with the frequency. Thus, the MC simulation enables the additional power WCS components to be significant.

The difference between significant level of Ge and the MC simulations is most likely caused by the heteroscedastic character of an input data, when in the time period of commoved series are the sub-periods with the different variances (you can see Fig. 1). An additional source of the difference between the significance level could be the scale range selection. Therefore, it is suitable to evaluate and assess obtained results in the context of application area (economy etc.).

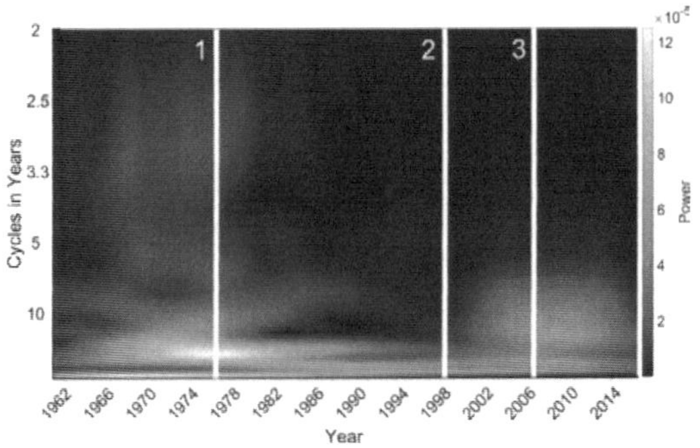

Fig. 3. Selection of slices in co-movement in three different time moments (1: 1976, 2: 1998, 3: 2006)

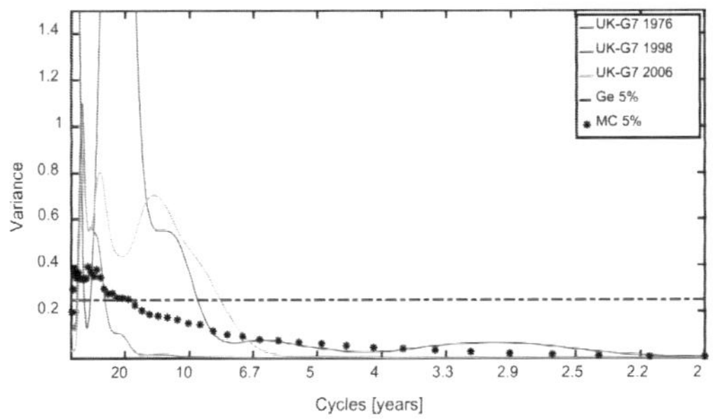

Fig. 4. Comparison of Ge and MC significance level

4 Conclusion

Using real data we analysed the behaviour of testing approach for the estimated wavelet power cross spectrum with respect to the GWN background. We use two approaches, namely the statistical significance according Ge, and the identification of critical value using MC simulations. While Ge uses comparison with constant critical value identified via the modified Bessel function of order zero corresponding to the risk, the MC simulation identify critical value on the basis of repeated random sampling to identify empirical distribution.

On the basis of obtained results we suggest using MC simulations to identify critical values for cross spectrum rather than Ge approach especially when the data are heteroscedastic. It means in the case, when the data have the different variance during its time evolution. In such case, the Ge approach cannot be able to capture all significant commoved areas. As supporting step we see as suitable to assess obtained results in the context of the scientific area from which the data originate.

The research described in the paper was supported by the Czech Science Foundation via grant n. 17-24309S and by the Czech Ministry of Education in the frame of National

Sustainability Program under grant LO1401. For research, the infrastructure of the SIX Center was used.

References

1. I. Korhonen, A. Peresetsky, What influences stock market behavior in Russia and other emerging countries, *Emergin Market Finance & Trade*, **52** (2016)
2. J. Janku, S. Kappel, Z. Kucerova. Monetary and Fiscal Policy Coordination in Slovakia: A Game Theory Approach, *Ekonomický časopis* **63** (2015)
3. S. Kapounek, Z. Kučerová, J. Fidrmuc., Lending conditions in EU: The role of credit demand and supply, *Economic Modelling* (2017)
4. G. Hassan, A. Cooray, M. Holmes, The effect of female and male health on economic growth: cross-country evidence within a production function Framework, *Empirical Economics*, **52** (2017)
5. M. Feldkircher, I. Korhonen, The rise of China and james charles its implications for the global economy:evidence from a global vector autopregressive model. *Pacific Economic Review*, **19** (2014)
6. E. Lukmanova, G. Tondl, Macroeconomic imbalances and business cycle synchronization. Why common economic governance is imperative for the Eurozone. *Economic Modelling* **62** (2017
7. F. Lu, H. Qiao, S. Wang, K. K. Lai, Y. Li, Time-varying coefficient vector autoregressions model based on dynamic correlation with an application to crude oil and stock markets. *Environmental Research* **152** (2017)
8. Ch. Croux, M. Forni, L. Reichlin, A Measure of Comovement for Economic Variables: Theory And Empirics. *The Review of Economics and Statistics* **83** (2001)
9. A. Iacobucci, A. Noullez, A frequency selective filter for short-length time series. *Computational Economics* **25** (2005)
10. J. Fidrmuc, T. Ikeda, K. Iwatsubo, International transmission of business cycles: Evidence from dynamic correlations. *Economic Letters* **114** (2012)
11. J. Poměnková, S. Kapounek, R. Maršálek, Variability of Dynamic Correlation–The Evidence of Sector-Specific Shocks in V4 Countries. *Prague Economic Papers* **23** (2014)
12. A. Xu, S. Haykin, R. J. Racine, Multiple Window Time-Frequency Distribution and Coherence of EEG Using Slepian Sequences and Hermite Function. *IEEE Transactions of Biomedical Engineering* **46** (1999)
13. J. G. Proakis, Ch. M. Rader, F. L. Ling, Ch. L. Nikias, M. Moonen, J. K. Proudler. 2002. *Algorithms for Statistical Signal Processing*. (Prentice Hall. 2002).
14. D. Wang, S. Shilong, W. Tse Peter., A general sequential Monte Carlo method based optimal wavelet filter: A Bayesian approach for extracting bearing fault features. *Mechanical Systems and Signal Processing* **52** (2015)
15. W. Jiang, S. Mahadevan, Wavelet spectrum analysis approach to model validation of dynamic systems. *Mechanical Systems and Signal Processing* **25** (2011)
16. R. Maršálek, J. Poměnková, S. Kapounek. A Wavelet-Based Approach to Filter Out Symmetric Macroeconomic Shocks. *Comput. Econ.* **44** (2014)
17. E. Klejmová, T. Malach, J. Poměnková, Wavelet Significance Testing with Respect to GWN Background: Monte Carlo Simulation Usage. In *Proceedings of 27th international conference Radioelektronika 2017.*
18. E. Klejmová, J. Poměnková, J. Blumenstein, Combination of the time-frequency representation for background noise supression. In *Proceedings 31st European Conference on Moelling and Simulation ECMS* 2017.
19. Z. Ftiti, A. Tiwari, A. Belanés, Tests of Financial Market Contagion: Evolutionary Cospectral Analysis V. S. Wavelet Analysis. *Computational Economics* **46** (2014)
20. A. N. Berdiev, Ch-P. Chang, Business cycle synchronization in Asia-Pacific: New evidence from wavelet analysis. *Journal of Asia Econ.* **37** (2015)
21. Ch. Torrence, G. P. Compo, A practical guide to wavelet analysis. *Bulletin of the American Meteorological society* **79** (1998)
22. Z. Ge, Significance tests for the wavelet cross spectrum and wavelet linear coherence. *Annales Geophysicae* **26** (2008)
23. Z. Ge, Significance tests for the wavelet power and the wavelet power spectrum. *Annales Geophysicae* **25** (2007)
24. W. T. Wells,R. L. Anderson, J. W. Cell, The distribution of the productof two central or non-central chi-square varietes, *Ann. Math. Stat.* **33** (1962)
25. C. P. Robert, G. Casella, *Monte Carlo Statistical Methods* (New York: Springer, 2004)
26. Organisation for Economic Co-operation and Development: National Accounts [online database]. (2017) [cit. 2017-01-12]. Available at: http:$\backslash\backslash$stats.oecd.org$\\ \backslash$Index.aspx?DatasetCode$=$SNA_TAB LE1.

Effect of DCT Image Compression on Eye Gaze Direction Detection

Kamil Dimililer[1,*], *Yoney* Kirsal Ever[2], *Cansu* Somturk[1], *Fulden* Ergun[3], *Guner* Urun[1], and *Mehmet* Kara[1]

[1]Electric and Electronic Engineering Department, Near East University, North Cyprus, Mersin 10 Turkey
[2]Software Engineering Department, Near East University, North Cyprus, Mersin 10 Turkey
[3]Computer Engineering Department, Near East University, North Cyprus, Mersin 10 Turkey

Abstract. The prediction of eye direction detection is the one of the popular research topic in human computer interaction area. This paper defines eye gaze detection by using Discrete Cosine Transform. Actually, determining the position of eyes is difficult to estimate the location of gaze which is more challenging. The database of the suggested research is organized as gaze directions of right, left and centre. The database is created with varied ages of images. In this paper, Discrete Cosine Transform has been applied to the image database and the effect of image compression is tested by using back propagation neural networks.

1 Introduction

In human body, every organ has its own importance, however eyes provide communication with the world [1]. In researcher's work area, the direction of eye is estimated and classified by using different techniques in recent years. With the development of techniques, like virtual reality and human computer interaction (HCI), gaze tracking support that it is very useful in practical applications [2]. Especially, monitoring of eye-gaze movements is frequently studied on HCI, usability testing, visual systems, psychology and many other areas. It is mainly used to create an alternative user interface in the field of HCI [3]. The process of the finding gaze direction classification is to detect eye in the images first and then to process the obtained information by using various types of implementations. The point of view of the eye is a good display to attract attention and attention of a person [4]. In recent years, the nature of eye movements has been considered like a biometric pattern in some HCI researches [5-6].

Basically, three different approaches are used for monitoring eye gaze and eye movements detection. These are user views from normal sensor cameras using image processing techniques on image based methods or to the skin region around the eye based on the measurement of electric field with sensors installed with electro holographic methods or infra-red light source [7]. The most commonly used method is the first method. In operation, the display will eye gaze monitoring has been performed. Figure 1 represents sample input images from the database.

When the system takes selected input images, the function of face detection will be activated [8]. Then eyes and mouth positions was detected. However, detection of eye gazes is a challenging topic [9-10]. Purpose of this research is just focusing on obtaining real gaze positioning.

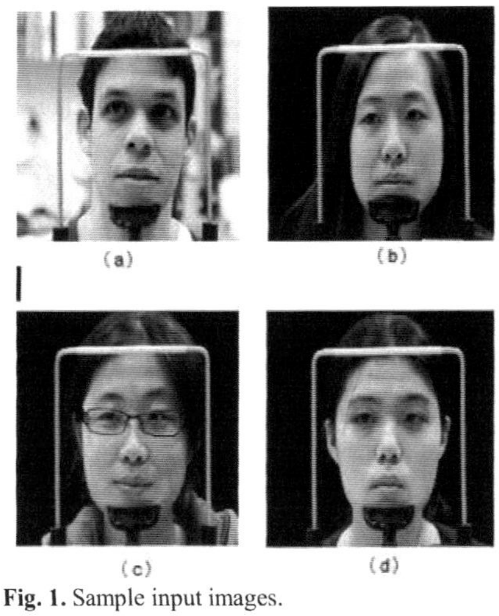

Fig. 1. Sample input images.

Image processing and neural networks are frequently used in detection purposes in different fields of the daily lives [21-22]. This paper represents the way of eye detection and direction of gaze classification. The research database is organized according to the direction of gaze images considering the three directions, right, left as well as centred. All database images are converted to grayscale and image resizing has been applied following the Discrete Cosine Transform (DCT) in order to prepare the database for the input of neural network [20, 23].

* Corresponding author: kamil.dimililer@neu.edu.tr

2 Related Works

In literature, the classification of gaze of eye direction are based on different/numerous approaches [11]. In this research [11], classification techniques are also used to estimate eye detection and gaze direction. As stated earlier, different techniques such as Electrocardiogram (EOG) are used by the placed electrodes to measure electric potentials around the eyes for estimating and analysing eye gaze [12]. In this EOG method, eyes are needed to be in original position, and the electrodes are used to detect the electric potential field. The movements are observed for approximate calculation of the eye location by measuring EOG signal [12].

Lui et al. [13], proposed a real-time system on detecting eyes. In this system, a detector for face and Template Matching (TM) are used together for detecting the face and locating the eyes. Additionally, Zernike Moments (ZM) are applied to extract rotation invariant eye features, and Support Vector Machine (SVM) is considered for the eye/non-eye patterns classification. Eye positions are estimated by selecting the maximum values for left and right positions in the eye probability map.

Recently another approach using component separation to detect eye direction is proposed by Vranceanu et al. [14]. A Machine Learning framework is proposed to classify the gaze direction for Iris, sclera, and skin. Additionally, gaze estimation is calculated by using Convolutional Neural Network (CNN) [15]. They combined face pose estimator's data with eye region. In the output layer of the CNN model is trained as a regression model.

The remaining part of this paper is organized as follows. Section III describes the proposed approach for system. An approach for feature extraction and an encoding is described in Section IV. The proposed novel matching schemes by SVM and ANN are presented in Section V and VI respectively. The experimental results and discussions are shown in Section IV.

3 Proposed Approach for System

This section is divided into two. In image processing phase, DCT compression has been applied to the eye database, and in second phase, compressed image set is applied for detecting direction of eye gaze, by using a 3-layer back propagation neural network (BPNN) using an image size of 256 x 256.

3.1 Discrete Cosine Transform (DCT)

Discrete Cosine Transform (DCT) exploits cosine functions, it transforms a signal from spatial representation into frequency domain. The DCT represents an image as a sum of sinusoids of varying magnitudes and frequencies. DCT is an example of transform coding.

The Joint Photographic Expert Group (JPEG) system, based on DCT, has been the most widely used compression method.

3.2 Back Propagation Neural Networks (BPNN)

A back propagation neural networks (BPNN) with 65536 input neurons is created. Output neurons classify the existing images as left, right and middle gaze. In order to activate neurons in the hidden and output layers, the sigmoid activation function is used.

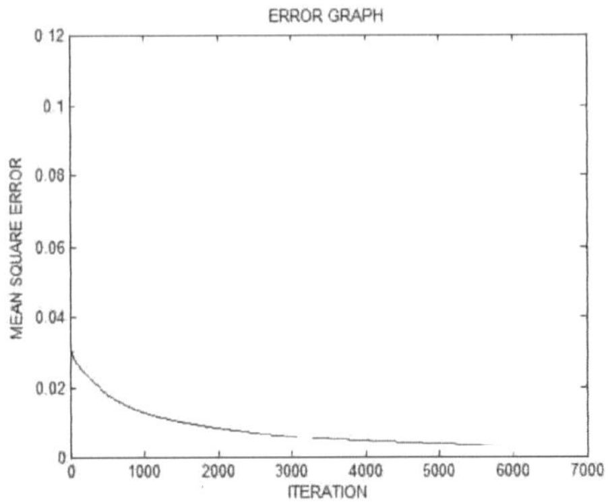

Fig. 2. BPNN Curve Graph.

Figure 3 shows the topology of the suggested BPNN that is a supervised learner algorithm. Since the implementation of BPNN is simple, and the availability of sufficient "input – target" database for training have been preferred.

In this research phase's learning or training and generalization are comprised. The organization of image database is shown below:

- Learning set: 60 images
- Generalization set: 90 images

Values between -0.4 and 0.4 were used in the learning state for initial random weights. The learning rate as well as the momentum factor has been chosen experimentally to achieve the required minimum error value and meaningful learning. 0.002 was considered as sufficient error value for this application. Table 1 shows the final parameters.

4 Results and Discussions

Considering the proposed application, similar to existing studies [16-19], results of the system are obtained from a virtual machine that has a configuration of 2012R2 64 bits OS. It is connected to a super computer with 1 Gbits speed. The super computer's specifications are listed as IBM Blade HS21 XM with 1280 Intel Xeon core units of 2.33 GHz CPU. Intelligent eye gaze detection system has been demonstrated with the robustness, flexibility and speed. Eye gaze identification results that use the training image set resulted 100% recognition ratio as expected.

The image set for testing of the proposed BPNN results were prosperous and reassuring. An overall correct identification of BPNN resulted 84.4% where, 76 out of 90 images yielded. This successful result was obtained by using only the database of images for training the neural network.

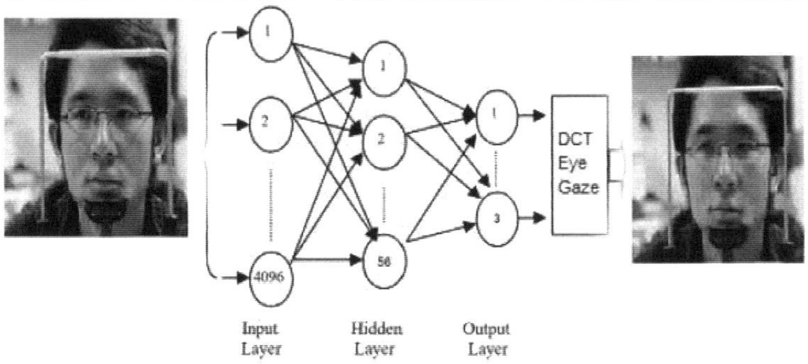

Fig. 3. BPNN Architecture.

Table 1. BPNN Training final parameters.

Input layer	65536
Hidden Layer	56
Output Layer	3
Learning Coefficient	0.005
Momentum Factor	0.5
Minimum RMS Error	0.002
Number of Iterations	5835
Training Time (Sec.)	95

5 Conclusion and Future Works

In this research, image processing and neural network approaches for the detection of eye gaze were developed. Image processing techniques are very useful and significant in many application fields.

The developed DCT system to notice the presence of gaze of the eye directly by the first look.

The system is based on an algorithm that reads an image, and performs DCT.

The images used in testing the proposed system are collected from database to get a total of 150 images for 60 images for training (20 images for left, middle and right), and 90 images for testing where 30 images for left, middle and right.

The experimental results obtained when testing the proposed system proved that the developed eye gaze detection system is a robust image processing system that is capable of detecting eye gaze direction in particularly right, left and middle. Also, a back propagation neural networks, named as BPNN is created. The developed BPNN uses original images with size of 256 x 256, in order to detect eye gaze, whereas 0.002 seconds is needed for detection.

When DCT algorithm is considered, a recognition ratio of 84.4% has been reached with 76 images out of 90. When BPNN is considered, a recognition ratio of 92% has been reached, using the original image set.

Future work will include the effect of image compression using a 64 x 64 size of images, and also, image enhancement will be applied in pre-processing phase.

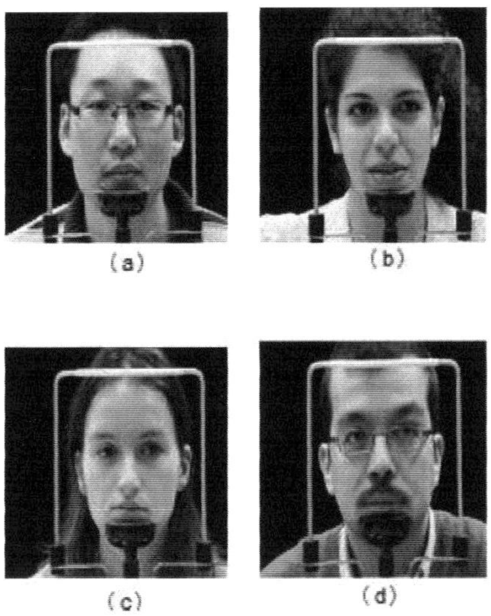

Fig. 4. Output images.

References

1. D. Beymer and M. Flickner, Eye Gaze Tracking Using an Active Stereo Head, Proceedings of the IEEE Computer Society Conference on Computer Vision and Pattern (CVPR '03), **2**: 451-458, (2003).

2. A. Fathi, Y. Li, and J. M. Rehg, Learning to recognize daily actions using gaze, in Proc. Eur. Conf. Comput. Vis., 314–327, (2012).

3. P.M., Corcoran, F., Nanu, S., Petrescu, and P., Bigioi, Real-time eye gaze tracking for gaming design and consumer electronics systems,

Consumer Electronics, IEEE Transactions on, **8:** 347 –355, (2011).

4. K. R. Park, Real-Time Gaze Detection via Neural Network, ICONIP 2004, LNCS 3316, 673-678, (2004).

5. A. George and A. Routray, A score level fusion method for eye movement biometrics, Pattern Recognition Letters, (2015).

6. C. D. Holland and O. V. Komogortsev, Complex eye movement pattern biometrics: Analyzing fixations and saccades, in Biometrics (ICB), 2013 International Conference on. IEEE, 1–8, (2013).

7. H. R. Chennamma, and X., Yuan, A Survey on Eye-Gaze Tracking Techniques, Indian Journal of Computer Science and Engineering, **4:** 388-393, (2013).

8. B. Fasel and J. Luettin, Automatic Facial Expression Analysis: A Survey, Pattern Recognition, **36**: 259-275, (2003).

9. A. Kendon, *Gesture: Visible Action as Utterance.* Cambridge, U.K: Cambridge Univ. Press, (2004).

10. D. Hansen and Q. Ji, In the eye of the beholder: A survey of models for eyes and gaze, IEEE Trans. Pattern Anal. Mach. Intell., **32:** 478–500, (2010).

11. D. H. Yoo, J. H. Kim, B. R. Lee and M. J. Chung, Non-Contact Eye Gaze Tracking System by Mapping of Corneal Reflections, Proc. Fifth IEEE International Conference on Automatic Face and Gesture Recognition, 94-99, (2002).

12. J. Gips, P. Di-Mattia, F.X. Curran and P. Olivieri, Using Eagle Eyes: An Electrodes based Device for Controlling the Computer with Your Eyes to Help People with Special Needs, Proc. Interdisciplinary Aspects on Computers Helping People with Special Needs, 77-84, (1996).

13. H. Liu and Q. Liu, Robust Real-Time Eye Detection and Tracking for Rotated Facial Images under Complex Conditions, Sixth International Conference on Natural Computation (ICNC), **4**: 2028-2034, (2010).

14. R. Vr˘anceanu, C. Florea, L. Florea, and C. Vertan, Gaze direction estimation by component separation for recognition of eye accessing cues, Machine Vision and Applications, **26**: 267–278, (2015).

15. X. Zhang, Y. Sugano, M. Fritz, and A. Bulling, Appearance-based gaze estimation in the wild, in 2015 IEEE Conference on Computer Vision and Pattern Recognition. IEEE Computer Society, (2015).

16. K. Dimililer, Back-propagation neural network implementation for medical image compression. J Appl Math, (2013).

17. K. Dimililer, Neural network implementation for image compression of x-rays. Electronics World, **118**:26-29, 2012.

18. K. Dimililer, Y. Kirsal Ever, and H. Ratemi, Intelligent eye tumour detection system, Procedia Computer Science, **102**:325-332, (2016).

19. K. Dimililer, Y. Kirsal Ever, and B. Ugur, ILTDS: Intelligent lung tumor detection system on ct images. Intelligent Systems Technologies and Applications, 225-235, (2016).

20. K. Dimililer, A. Hesri and Y. Kirsal Ever, Lung Lesion Segmentation using Gaussian Filter and Discrete Wavelet Transform, The Journal of ITM Web of Conferences, EDP Sciences, ISSN 2271-2097, **11**, (2017).

21. I. J. Bush and K. Dimililer, Static and Dynamic Pedestrian Detection Algorithm for Visual Based Driver Assistive System, The Journal of ITM Web of Conferences, EDP Sciences, ISSN 2271-2097, **9**, (2017).

22. K. Dimililer, A. Hesri and Y. Kirsal Ever, Lung Lesion Segmentation using Gaussian Filter and Discrete Wavelet Transform, The 2017 International Conference on Information Science and Technology, **11**, (2017).

23. I. J. Bush and K. Dimililer, Static and Dynamic Pedestrian Detection Algorithm for Visual Based Driver Assistive System, The 2016 International Conference Applied Maths, Computational Science and Systems Engineering (AMCSE2016), **9**, (2017).

Planning pesticides usage for herbal and animal pests based on intelligent classification system with image processing and neural networks

Kamil Dimililer[1,*], *Yoney* Kirsal Ever[2]
[1]Electric and Electronic Engineering Department, Near East University, North Cyprus, Mersin 10 Turkey
[2]Software Engineering Department, Near East University, North Cyprus, Mersin 10 Turkey

Abstract. Pests are divided into two as herbal and animal pests in agriculture, and detection and use of minimum pesticides are quite challenging task. Last three decades, researchers have been improving their studies on these manners. Therefore, effective, efficient, and as well as intelligent systems are designed and modelled. In this paper, an intelligent classification system is designed for detecting pests as herbal or animal to use of proper pesticides accordingly. The designed system suggests two main stages. Firstly, images are processed using different image processing techniques that images have specific distinguishing geometric patterns. The second stage is neural network phase for classification. A backpropagation neural network is used for training and testing with processed images. System is tested, and experiment results show efficiency and effective classification rate. Autonomy and time efficiency within the pesticide usage are also discussed.

1 Introduction

Last three decades, people have been concerned about health and safety of the food consuming. Since there are several hundreds of pest species in paddy fields, various types of pesticides are applied but these health and safety awareness make the agricultural staff to minimize the use of pesticides in necessary amount. In order to decide the required pesticide, visual inspection and comparison of insects is commonly used method around the world. This approach is highly challenging and requires considerable training and experience with insects. The experimental methods are time-consuming, affordable, and not suitable for routine use [27].

Recently, an increment in the usage of image processing techniques in various applications as pre-processing stage are seen [29-31]. Image processing techniques have been applied in different ways to identify crops, plants, leaves as well as diseases in agriculture paddy fields [1]. According to their research, an image segmentation technique is used for detection of agricultural products. In [24] paper, a short survey for potential future techniques for image processing in different agricultural areas were highlighted to guide researchers and farmers. [23] did a similar study to increase the productivity while reducing the diseases. In another image processing research, a software was developed to decide different vegetation indexes and the 3-dimensional feature of crops by [20].

Artificial Neural Networks is widely used in the classification of different objects [30, 31]. In order to classify pests, biometric based features are adopted for automatically identifying and classifying the images of insects. The primary advantage of this approach is that it is invariant to changes in pose and scale. Furthermore, to achieve a high classification accuracy, a backpropagation neural classifier is required for detection.

The main aim of this research is to classify different types of pests based on their geometric features that are extracted using edge detection. The classification is performed using a backpropagation neural network due to its simplicity and efficiency in such applications [10]. Different image processing techniques are applied to images in order to highlight the edges and prepare them for the back propagation neural networks. The images are prepared for simulation and testing purposes [7]. Therefore, three backpropagation neural networks are used. Two types of edge detection methods have been applied to the available image set after applying the median filtering in order to extract the features of the insects. The prepared image set will be prepared for the input layer of the Back Propagation Neural Networks, BPNN. The simulation results show differences in the classification rates and effectiveness of the suggested networks.

The experiment demonstrates a classification success rate of 84% for the first network, 90% for the second network and 92% for the third network. The

[*] Corresponding author: kamil.dimililer@neu.edu.tr

method developed in this paper offers potential for the classification of pests.

2 Related Works

In literature, there are different types of neural network algorithms used for categorization or classification of various images [15]. In this paper, the proposed system is divided into two categories; image processing stage and neural networks stage. The images are classified as herbal and animal pests such as Agrotis Ipsilon, Gryllotalpa, carponya, tetrancyhidae, Mamestra, Lertinotarsa, Siphonaptera, ixodida, Siphonaptera, Cimicidae, Phlebotomus, Aphidoidea. The image processing phase consists of grayscale conversion, filtering, Sobel and Canny edge detection. These techniques are used to highlight the geometric shapes of the pests taking into account the distinguished features. At the end of this phase, the images are ready to be fed into the proposed neural network phase.

2.1 Dataset

Due to the unavailability of a single source of insect images, the images were collected from different websites for insect classification. Resultantly, the total number of images was 120, out of which 10 were Agrotis, 10 were Gryllotalpa, 10 were Carpomya, 10 were Tetranychidae, 10 were Mamestra, 10 were Lepinotarsa, 10 were Siphonaptera, 10 were ixodida, 10 were Siphonaptera, 10 were Cimicidae, 10 were Phlebotomus. 10 were Aphidoidea. Figure 1. Shows the sample database images.

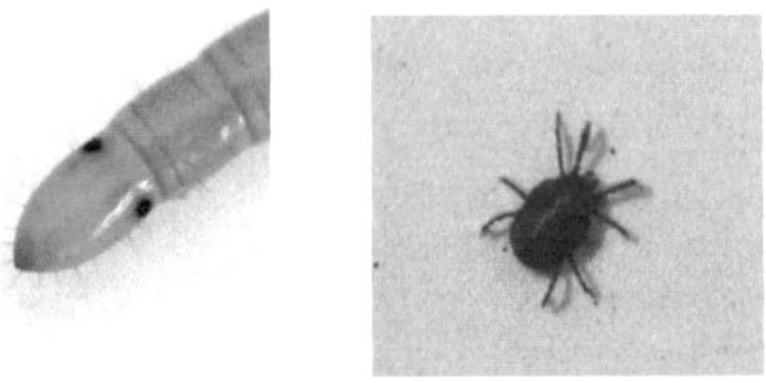

(a) Herbal Pests

(b) Animal Pests

Fig. 1. Sample images from the database.

3 Proposed Systems

This section focusses on utilities and parameters of the proposed system. Two phases, image processing and neural network phases are designed respectively.

3.1 Image Processing Phase

A bunch of image processing methods is applied to the pest database for acceptable amount of identification.

3.1.1 Formatting the title

In order to minimize the CPU Time, the original RGB images are converted to grayscale. Using the luminosity method, grayscale images are more precise [3]. Figure 2 shows the conversion of a colour image into a grayscale image.

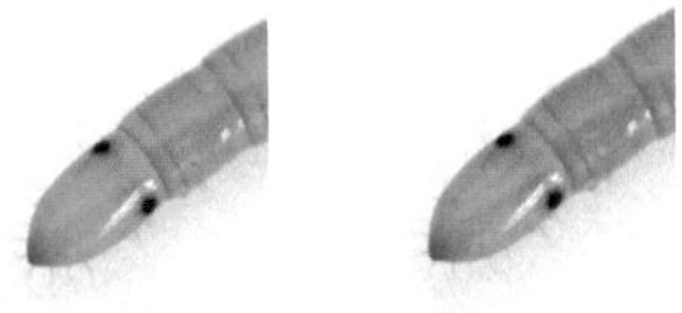

(a) Herbal Pests

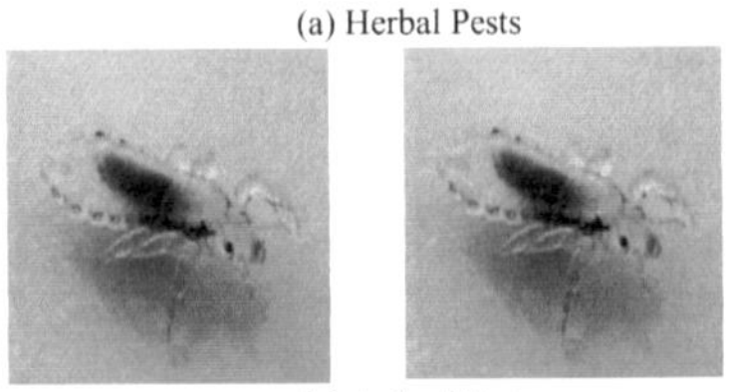

(b) Animal Pests

Fig. 2. RGB to Grayscale conversion.

3.1.2 Image smoothing with filtering

Smoothing is one of the image processing techniques used for reducing the details within an image in order to create a clearer image [3]. The most common type of linear filter is the median filter which is used to reduce impulsive noise [6]. The median filter goes through each element of the image and replaces each pixel with the median of its neighboring pixels that are located in a square neighborhood (kernel) around the evaluated pixel [25]. Figure 3 represents the application of image filtering to the pest insects.

(a) Herbal Pests

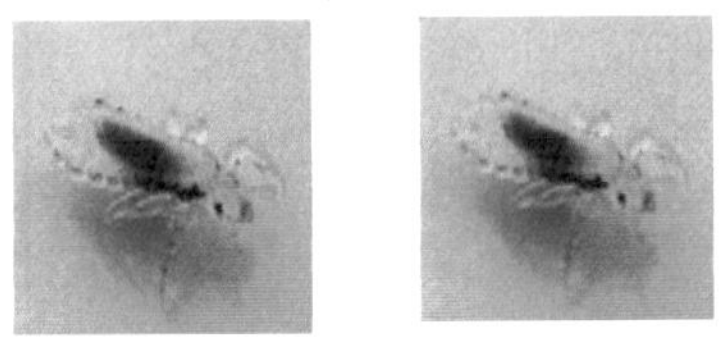

(b) Animal Pests

Fig. 3. Filtering applied images.

3.1.3 Thresholding

Image thresholding separates a particular region of image into two regions. First region is the region for the foreground, where second one is the background region for the unneeded objects. In proposed system, the input data is a grayscale image, where the output is a binary image that [0] (black pixels) represents background and [1] (white pixels) represents foreground. Intensity threshold is set by analyzing the histogram of the image.

This represents the intensity distribution of that particular image [10]. Figure 4 represents the application of thresholding to the filtered grayscale image.

(a) Herbal Pests

(b) Animal Pests

Fig. 4. Threshold applied images.

3.1.4 Edge detection based segmentation

Segmentation is defined as grouping of required regions of the image into foreground and background regions of interest and removed data respectively [8].

Edge detection is identified as sharp intensity contrasts in an image. In this proposed system, edges are detected using Canny and Sobel operators. Canny edge detection technique provides good noise immunity and detects the through edges while minimizing errors. On the other hand, Sobel edge detection technique uses 3-by-3 kernels which calculates approximation of the horizontal and vertical derivatives [2,14]. Figure 5 shows the segmentation applied images using canny and sobel operators.

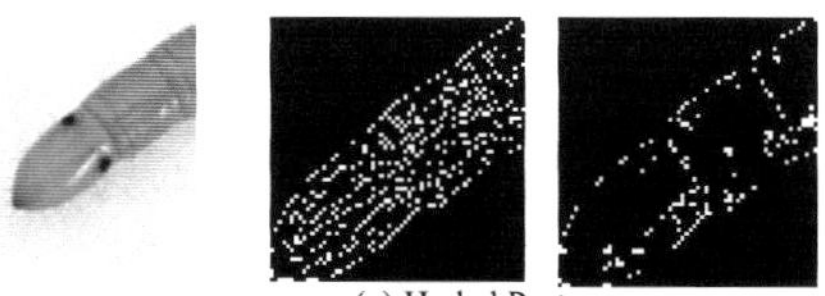

(a) Herbal Pests

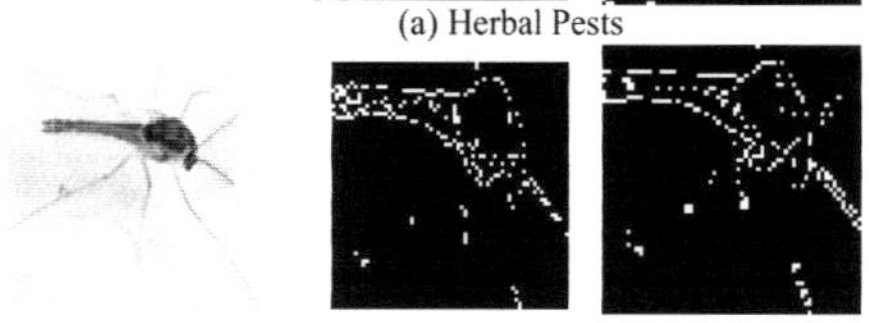

(b) Animal Pests

Fig. 5. Segmentation applied images.

3.2 Neural Networks Phase

In this phase, the images of the different pests were classified into two classes using a supervised backpropagation neural classifier. One database was used for training and testing the different networks. However, three learning schemes were used [4, 5].

The proposed system consists of a database of 60 images for herbal pests and 60 images for animal pests where 70 images are used for training purpose whereas 50 images were used for testing. There are 3 BPNN Learning schemes used to test the efficiency of the suggested application. 65536 pixels represents the 256x256 grayscale image that is used as the input data.

Figure 6 represents the 3 topologies of the suggested neural network phases where Figure 6(a) shows the original grayscale images as inputs, Figure 6(b) shows the Canny edge detection applied images as inputs, and Figure 6(c) shows the Sobel edge detection applied images as inputs.

Table 1. BPNN Training final parameters.

	BPNN1	BPNN2	BPNN3
	(Fig.6(a))	(Fig.6(b))	(Fig.6(c))
Input layer nodes	65536	65536	65536
Output layer nodes	2	2	2
Learning rate	0.004	0.004	0.003
Momentum rate	0.3	0.5	0.5
Minimum error	0.002	0.002	0.002

4 Results and Discussions

These results were obtained using a configuration of 2012R2 64 bits OS Virtual machine that has a 1 Gbits connection speed to a super computer that has IBM Blade HS21 XM Intel Xeon 2.33 GHz CPU with 1280 core units. Through this application, the robustness, flexibility and speed of this pest detection system have been demonstrated. Pests identification results using the training image set yielded 100% recognition as would be expected in BPNN.

Figure 6 shows the topologies of the suggested BPNN Systems. Due to the implementation simplicity, and the availability of sufficient "input – target" database for training, the use of a back propagation neural network which is a supervised learner, has been preferred.

Training and testing (generalization) are comprised for these phases. The available brain image database is organized as follows:

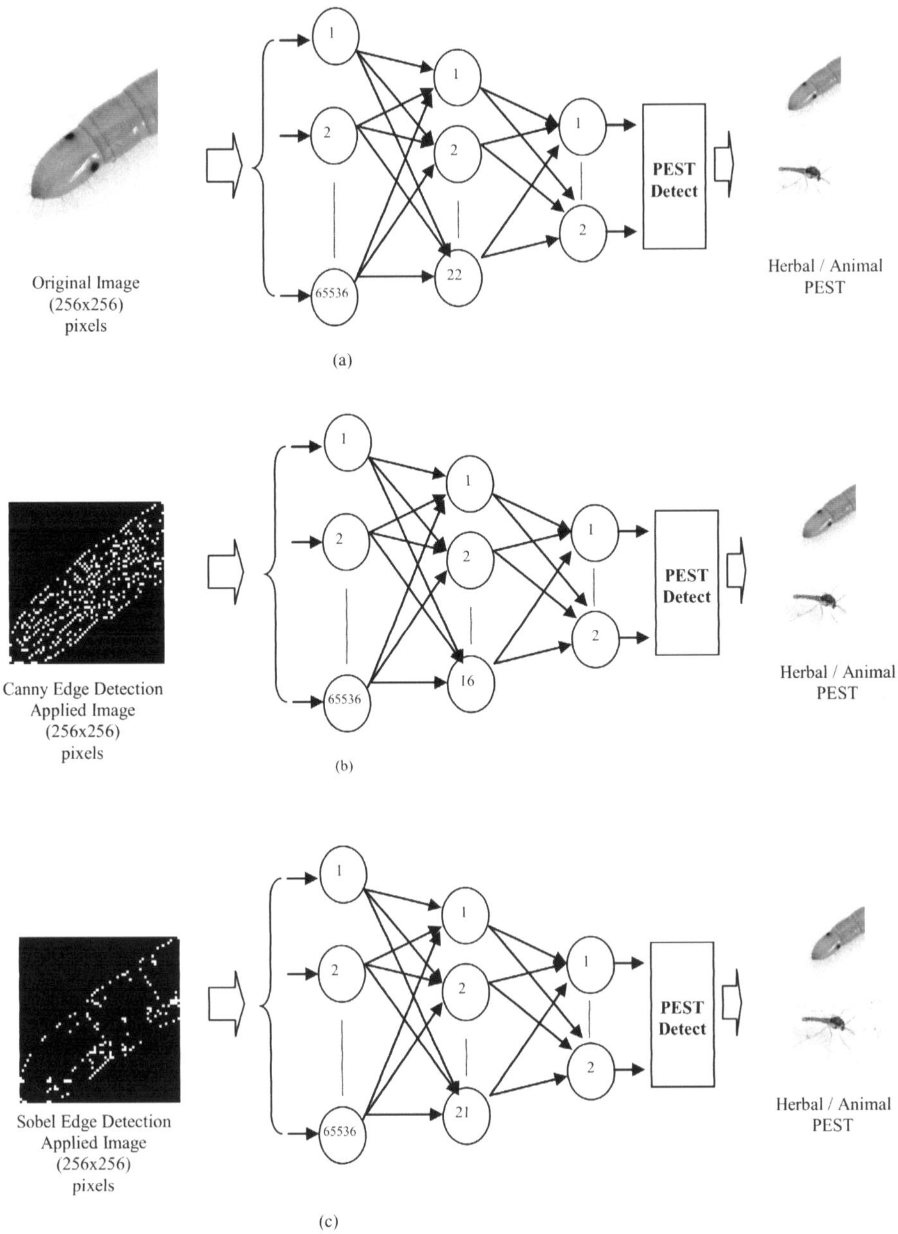

Fig.6. BPNN Architectures.

1. Training image set: 70 images
2. Testing image set: 50 images

During the learning phase, initial random weights of values between -0.3 and 0.3 were used. In order to achieve the required minimum error value and meaningful learning, the learning rate and the momentum rate were adjusted during various experiments.

In BPNN1, the identification of the images is yielded 84% where, 42 images out of the available 50 pest images correctly yielded, in BPNN2, the identification of the images is yielded 90% where, 45 images out of the available 50 pest images correctly yielded whereas in BPNN3, the identification of the images is yielded 92% where, 46 images out of the available 50 pest images correctly yielded.

As it can be seen from the results of the iterations, results were successful and encouraging. BPNN3 which represents Sobel edge detection algorithm yields optimum results, after several experiments. This successful result was obtained by using only the database of images for training the neural network.

5 Conclusion and Future Works

In this paper, image processing and neural network approaches for the detection of pests were developed. Image processing techniques are very useful and significant in various application fields.

The proposed system uses three back propagation neural networks, named as BPNN1, BPNN2, BPNN3. The developed BPNNs use pre-processed, original images to detect pests with and without fractures respectively.

The motivation of this work was to compare and evaluate the obtained results and also, effectiveness of a back propagation neural network in recognizing different herbal and animal pests. Comparing the identification rations as discussed in the results and discussions section, BPNN3 that is Sobel edge identification.

The developed system is a robust system since it was trained to recognize pests. The images used for training and testing the proposed system were collected from the benchmark database, which contains 120 images. The images contain different features in size and illumination conditions for each subject, which was an advantage for us to make the system more efficient and robust. Finally, it can be stated that the experimental analysis of the proposed pests detection system showed a great efficiency and an outperforming rate.

Last but not the least, the proposed BPNN algorithms results will guide architectural staff in term of detection and determination of the most efficient and least harmful pesticides.

References

1. R. Chahar and P. Soni A study of image processing in agriculture for detect the plant diseases. International Journal of Computer Science and Mobile Computing. **4**:581-587, (2015).

2. L., Chao, Z., Ji-liu and H.E., Kun, Adaptive edge-detection method based on Canny algorithm. Computer Engineering and Design, **31**:4036-4039, (2010).

3. J., Church, Y., Chen, and S., Rice, A Spatial median filter for noise removal in digital images, IEEE Southeast Conf., 618 – 623, (2008).

4. K. Dimililer, Backpropagation neural network implementation for medical image compression. J. Appl. Math., (2013).

5. K. Dimililer, Neural network implementation for image compression of x-rays. Electronics World. **118**:26-29, (2012).

6. K. Dimililer, and A. İlhan, Effect of image enhancement on mri brain images with neural networks, (2016).

7. K., Dimililer, Y. Kirsal Ever, and H. Ratemi, Intelligent eye tumour detection system. Procedia Computer Science, 102:325-332, (2016a).

8. K. Dimililer, Y. Kirsal Ever, and B. Ugur, ILTDS: Intelligent lung tumor detection system on ct images. Intelligent Systems Technologies and Applications, 225-235, (2016b).

9. K. Dimililer, Y. Kirsal Ever and B. Ugur, Tumor Detection on CT Lung Images using Image Enhancement, The Online Journal of Science and Technology, **7**, (2016c).

10. K. Dimililer and S. Zarrouk, ICSPI: Intelligent classification system of pest insects based on image processing and neural arbitration, American Society of Agricultural and Biological Engineers, **33**:4, (2017).

11. Y. K. Ever, E. Gemikonakli, and K. Dimililer, Modelling Approaches of Performance Evaluation Of High QoS Of Kerberos Server With Dynamically Renewing Keys Under Pseudo Conditions, The Online Journal of Science and Technology, **7**, (2016).

12. H.L, Eng and K.K, Ma, Noise adaptive soft-switching median filter. IEEE Trans Image Process, **10**:242-251, (2001).

13. Gonzalez and Woods, *Digital Image Processing*, Second Edition, Prentice Hall,108-120, (2002).

14. Z., Jian-jun, and L. Jing, A surface crack edge detection algorithm based on improved Sobel operator. Journal of hefei university of technology, **34**:845- 847,960, (2011).

15. A. Khashman and K. Dimililer, Medical radiographs compression using neural networks and haar wavelet, IEEE EUROCON'09, 1448-1453, (2009).

16. Y. Kim, Contrast enhancement using brightness preserving bi-histogram equalization, IEEE Trans. Consum. Electron. **43**:1–8, (1997).

17. N., Larios, Deng, H., Zhang, W., Sarpola, M., Yuen, J., Paasch, R., Moldenke, A., Lytle, D. A., Correa, S. R., Mortensen, E. N., Shapiro, L. G., and Dietteric, T.G., Automated insect identification through concatenated histograms of local appearance features: feature vector generation and region detection for deformable objects, Machine Vision and Applications, Springer, **19**:105-123, (2008).

18. T.Z. Lee, Edge detection analysis. Int. J. Comp. Sci., **5**:1-25, (2012).

19. I., Levner and H. Zhangm. Classification driven watershed segmentation, IEEE Transactions on image Processing, **16**, (2007).

20. A. F. J., Lopez, Salamanca, J. M., Medina, M. J. Q., and Perez, O. E. A. Crops diagnosis using digital image processing and precision agriculture technologies., INGE CUC. **11**:63-71, (2015).

21. A. Mittal and, S.K., Dubey Analysis of rheumatoid arthritis through image processing. IJCSI International Journal of Computer Science. **9**, (2012).

22. R.G., Mundada, and, V. V. Gohokar, Detection and classification of pests in greenhouse using image processing, IOSR Journal of Electronics and Communication Engineering. 5:57-63. Procedia Computer Science, 102:39-44, (2013).

23. J. K., Patil, and R. Kumar. Advances in image processing for detection of plant diseases, Journal of Advanced Bio information Applications and Research, 135-141, (2011).

24. L. P. Saxena and L. J. Armstrong, A survey of image processing techniques for agriculture. Proceedings of AFITA 2014, 406-418, (2014).

25. L. Shapiro and G, Stockman, Image Segmentation. (1st ed.), Prentice Hall, Washington, pp. 200-280, (2000)..

26. The nature pictures, available at: [http://www.naturespic.com/NewZealand/result_keyword.asp?imagekeywords=Insects%2Fspiders&offset=60].

27. L. G., Thiago, Souza, E.S., Mapa, K.S., and David M. Application of complex networks for automatic classification of damaging agents in soya bean leaflets. Proceeding of the 18th IEEE International Conference, 1065 – 1068, (2005).

28. Y., Wan, Chen, Q., and Zhang, B. M. Image enhancement based on equal area dualistic sub-image histogram equalization method. IEEE Trans. Consum. Electron.,**45**,68-75, (1999).

29. K. Dimililer, A. Hesri and Y. Kirsal Ever, Lung Lesion Segmentation using Gaussian Filter and Discrete Wavelet Transform, The Journal of ITM Web of Conferences, EDP Sciences, ISSN 2271-2097, **11**, (2017).

30. I. J. Bush and K. Dimililer, Static and Dynamic Pedestrian Detection Algorithm for Visual Based Driver Assistive System, The Journal of ITM Web of Conferences, EDP Sciences, ISSN 2271-2097, **9**, (2017).

31. K. Dimililer, A. Hesri and Y. Kirsal Ever, Lung Lesion Segmentation using Gaussian Filter and Discrete Wavelet Transform, The 2017 International Conference on Information Science and Technology, **11**, (2017).

32. I. J. Bush and K. Dimililer, Static and Dynamic Pedestrian Detection Algorithm for Visual Based Driver Assistive System, The 2016 International Conference Applied Maths, Computational Science and Systems Engineering (AMCSE2016), **9**, (2017).

On attracting sets in artificial networks: cross activation

Felix Sadyrbaev[1]

[1]Institute of Mathematics and Computer Science, University of Latvia, LV-1459 Rainis boul. 29, Riga, Latvia

Abstract. Mathematical models of artificial networks can be formulated in terms of dynamical systems describing the behaviour of a network over time. The interrelation between nodes (elements) of a network is encoded in the regulatory matrix. We consider a system of ordinary differential equations that describes in particular also genomic regulatory networks (GRN) and contains a sigmoidal function. The results are presented on attractors of such systems for a particular case of cross activation. The regulatory matrix is then of particular form consisting of unit entries everywhere except the main diagonal. We show that such a system can have not more than three critical points. At least *n–1* eigenvalues corresponding to any of the critical points are negative. An example for a particular choice of sigmoidal function is considered.

1 Introduction

Artificial neural networks can be used successfully to model dynamics of gene expression in a living cell [1]. Investigation of mathematical models for gene regulatory networks (GRNs) had become a challenging problem for mathematicians and biologists. The mathematical models based on differential equations can help to understand the organization within cells and their dynamics. Such models can be found in [1], [2], [3]. GRN therefore are networks of interacting over time elements. This interaction is organized so that the living organism is able to adapt to changes in the outside environment. Genes produce proteins that regulate other genes. A particular gene regulates a number of other genes and they, in turn, regulate the activity of many genes downstream. On the other hand, a particular gene is influenced and regulated by many genes upstream. [4].
All together they form a genomic regulatory network of elements turning each other on and off. There are attempts in the literature to transfer principles of self-organization in GRN to networks of different nature.
For instance, in telecommunication systems, where changes are rapid and unpredictable, one can construct an optimal virtual network topology (VNT) [2], [3] by establishing a set of lightpaths between nodes. To treat changing in time (fluctuating) traffic on a VNT, adaptive VNT control methods should be invented, which reconfigure VNTs according to traffic conditions on VNTs. To develope such methods, one way is to observe attractor selection in biological systems (GRN in particular) that adapt to unknown changes in their surrounding environments and recover their conditions.
The mathematical models for GRN [5-7] and telecommunication networks [2], [3] have similar features. In this article we consider the dynamical system model that appears in [1], [2], [3]. Our goal is to describe attractors for the dynamical system in a particular case. The system of differential equations contains the so called regulatory matrix W that is used to roughly describe the relations between elements of a modelled network. In order to deal with possibly simple objects it is proposed that entries of regulatory matrix W can be of only three kinds: -1; 0 and $+1$, that corresponds respectively to inhibition, no relation and activation. Properties of dynamical systems modelling the network strongly depend on the structure of matrix W. The study of related differential systems which are 1) nonlinear, 2) depend on parameters and 3) may be of arbitrary size, is a highly nontrivial task. Even in the simplest cases the analysis of such systems provide nontrivial results. For instance, in the work [8] the simplified system (for a network of two elements) of the form

$$\begin{cases} x_1' = \dfrac{1}{1+e^{-\mu(x_2-\theta)}} - x_1, \\ x_2' = \dfrac{1}{1+e^{-\mu(x_1-\theta)}} - x_2 \end{cases} \tag{1}$$

was considered. It was proved that this system can have up to three critical points depending on the values of parameters μ and θ. The characteristics of critical points were given. Later [9] it was shown that n-dimensional system with regulatory matrix

$$W = \begin{pmatrix} 0 & 1 & \vdots & 1 \\ 1 & 0 & \vdots & 1 \\ \dots & \dots & \dots & \dots \\ 1 & 1 & \vdots & 0 \end{pmatrix} \tag{2}$$

also can have up to three critical points of the form $(x, \dots, x)$ (lying on the bisectrix) of definite type. This case was called "cross activation". The types of critical

Corresponding author: felix@latnet.lv

points and their dependence on two parameters were revealed. In what follows we try to accomplish the same task for systems that contain arbitrary sigmoidal function in their description. By sigmoidal function $f(z)$ we call a monotonically increasing to the unity continuous function that is positive at $z=0$ and has exactly one point of inflexion (that separates convexity from concavity). The function $f(z)=(1+exp(-\mu z))^{-1}$ used in system (1) is of this kind but there are many other sigmoidal functions. Sigmoidal functions have typical *S*-shaped form. There is usually a parameter μ that regulates how steep is the graph of *S*-shaped function.

2 General information

We consider systems of the form

$$\frac{dx_i}{dt} = f(\sum W_{ij} x_j - \theta) - x_i, \qquad (3)$$

where f is a sigmoidal function, W is a regulatory matrix defined by (2) and θ is a parameter. Our goal is to describe attractors for such a system. In extended form the system (3) looks as

$$\begin{cases} x_1' = f(x_2 + x_3 + \ldots + x_n - \theta) - x_1, \\ \ldots \quad \ldots \quad \ldots \\ x_n' = f(x_1 + x_2 + \ldots + x_{n-1} - \theta) - x_n, \end{cases} \qquad (4)$$

where θ is a positive parameter. To find the critical points one has to consider the system

$$\begin{cases} x_1 = f(x_2 + x_3 + \ldots + x_n - \theta), \\ x_2 = f(x_1 + x_3 + \ldots + x_n - \theta), \\ \ldots \quad \ldots \quad \ldots \quad \ldots \quad \ldots \\ x_n = f(x_1 + x_2 + \ldots + x_{n-1} - \theta). \end{cases} \qquad (5)$$

Due to monotonicity of $f(z)$ the following is true.

Lemma 1. All critical points $(x_1,\ldots,x_n)$ lie on the diagonal of a unit n-dimensional cube, that is, $x_1=x_2=\ldots=x_n$.

Proof. Divide the second line in (5) by the first line. Suppose that $x_2 > x_1$. One gets

$$1 < \frac{x_2}{x_1} = \frac{f(x_1 + x_3 + \ldots + x_n - \theta]}{f(x_2 + x_3 + \ldots + x_n - \theta]} < 1.$$

The contradiction is obtained. Similarly the case $x_2 < x_1$ can be considered. Therefore $x_2 = x_1$. Considering the second and the third lines one can prove that $x_2 = x_3$. And so on. □

In view of the above lemma to find the critical points one has to solve the scalar equation

$$x=f((n-1)x-\theta). \qquad (6)$$

We thus have arrived at the statement.

Theorem 1. System (4) can have no more than three critical points.

The proof follows from the fact that the straight line $y=x$ cannot intersect S-shaped graph more than three times.

After detecting the critical points further analysis can be made in order to describe the attractors of the system.

We show the analysis of some particular case. The case of the logistic function $f(z)=(1+exp(-\mu z))^{-1}$ will be considered.

3 SYSTEM: GENE REGULATION

The system

$$\begin{cases} x_1' = \dfrac{1}{1+e^{-\mu(x_2+x_3+\ldots+x_n-\theta)}} - x_1, \\ \ldots \quad \ldots \quad \ldots \\ x_n' = \dfrac{1}{1+e^{-\mu(x_1+x_2+\ldots+x_{n-1}-\theta)}} - x_n, \end{cases} \qquad (7)$$

appears in the gene regulation theory [1], [5]. The parameters μ and θ are positive. In view of the results of the previous section all critical points are of the form $(x,\ldots,x)$ (n times), where x is to be determined from the relation

$$\theta = \frac{1}{\mu}\log(\frac{1}{x} - 1) + (n-1)x \qquad (8)$$

Analyzing this relation one obtains two branches

$$x_{1,2} = \tfrac{1}{2} \pm \sqrt{\frac{1}{4} - \frac{1}{(n-1)\mu}}.$$

Putting these branches into (7) two branches $\theta_1(\mu)$ and $\theta_2(\mu)$ can be defined that bound the region Ω in the (μ, θ) plane.

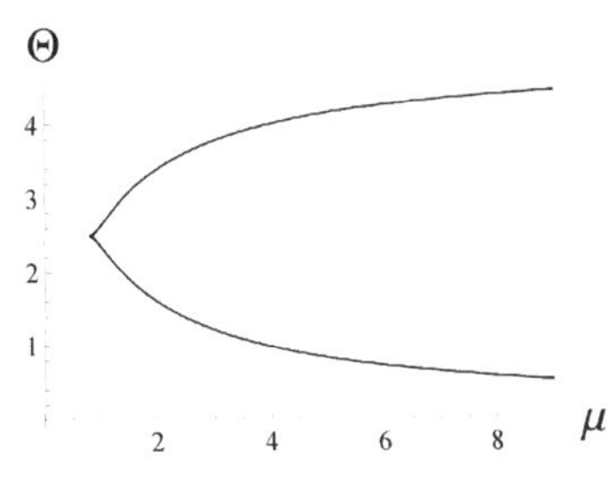

Fig. 1. Region Ω (bounded by two branches) for $n=6$.

3.1. Linearization

To detect types of critical points use the standard linearization procedure. The linearized at a critical point *(x, ... , x)* system is

$$\begin{cases} u_1' = -u_1 + \mu(1-x)x u_2 + \ldots + \mu(1-x)x u_n, \\ u_2' = -u_2 + \mu(1-x)x u_1 + \ldots + \mu(1-x)x u_n, \\ \ldots \ \ldots \ \ldots \ \ldots \ \ldots \\ u_n' = -u_n + \mu(1-x)x u_1 + \ldots + \mu(1-x)x u_{n-1}. \end{cases}$$

The characteristic values for the above linear system are to be determined from the equation

$$\begin{pmatrix} -1-\lambda & \mu x(1-x) & \vdots & \mu x(1-x) \\ \mu x(1-x) & -1-\lambda & \vdots & \mu x(1-x) \\ \ldots & \ldots & \ldots & \ldots \\ \mu x(1-x) & \mu x(1-x) & \vdots & -1-\lambda \end{pmatrix} = 0 \tag{7}$$

This equation, due to its specific structure, can be solved and the roots are

$$\lambda_1 = \ldots = \lambda_{n-1} = -\mu x(1-x) - 1, \\ \lambda_n = (n-1)\mu x(1-x) - 1 \tag{8}$$

Since for any possible critical point the first n-1 characteristic values λ_i are negative (recall that $x \in (0,1)$), the n-dimensional neighbourhood of any critical point contains (n-1)-dimensional attracting subspace. As to the remaining characteristic value λ_n (that determines the behaviour along the remaining *1*-dimensional subspace), it may be negative, zero or positive. Let us detect the sign of λ_n.

Lemma 2. If (μ, Θ) is in *ext* Ω, then the characteristic number λ_n is negative for a respective unique critical point;
if (μ, Θ) belongs to $\partial\Omega$, λ_n is negative for the first critical point and λ_n is zero for the second one, corresponding to an extremum of $\Theta(x)$;
if (μ, Θ) is in *int* Ω, λ_n is positive for one (in the middle) critical point and negative for two other (side) critical points; λ_n is zero also for a unique critical point corresponding to the vertex of Ω.

Theorem 2. An attracting set for system (7) consists of at most three critical points. The number of critical points and their character depend on the choice of parameters (η, θ). The following is true:

1) If $(\mu, \Theta) \in ext\ \Omega$, the attracting set consists of one critical point (attraction in all dimensions);

2) If $(\mu, \Theta) \in \partial\Omega$, the attracting set consists of two critical points, one attracting in all dimensions (all characteristic numbers are negative) and the second one attracting in (n-1)-dimensions and degenerate (the respective λ is zero) in the remaining dimension; the same (the respective λ is zero) is the character of the critical point corresponding to the vertex of region Ω;

3) if $(\mu, \Theta) \in int\ \Omega$, there are three critical points; the middle point is attracting in (n-1)-dimensions and repelling in the remaining dimension (the respective λ is positive); the two side critical points are attracting in all dimensions.

3.2. Conclusion

Systems of the form (4) arise in mathematical models of artificial networks including GRN networks. In the case of cross activation (the regulatory matrix consists of unit entries except the main diagonal) all critical points are on the bisectrix of a unit n-dimensional cube. The maximal number of critical points is three. At least n-1 eigenvalues (roots of the characteristic equations) for any critical point are negative. The cases of two and one critical points are possible.

References

1. J. Vohradský, FASEB J. **15**(3/2001), 846-54.
2. Y. Koizumi et al., Journal of Lightwave Technology, **8** (06/2010), Issue 11, 1720 – 1731.
3. Y. Koizumi, T. Miyamura, S. Arakawa, E. Oki, K. Shiomoto, and M. Murata, in Proceedings of BIONETICS, Nov. 2008, 1– 8.
4. A. Wuensche, Pac Symp Biocomput. 1998:89-102.
5. F. M. Alakwaa, Journal of Computational Systems Biology, **1**(1/2002), 67-103.
6. A. Crombach, P. Hogeweg, PLoS Comput Biol **4**(7), 2008: e1000112. doi:10.1371/journal.pcbi.1000112
7. A.Tušek and Ž. Kurtanjek, Ch. 5 in "*Applied Biological Engineering - Principles and Practice*" Ed. by Ganesh R. Naik, ISBN 978-953-51-0412-4, 674 pages, Publisher: InTech, DOI: 10.5772/2101
8. S. Atslega, D. Finaskins and F. Sadyrbaev, Mathematical Modelling and Analysis, **21** (3/ 2016), 385-398 DOI:10.3846/13926292.2016.1172131
9. E. Brokan and F. Sadyrbaev, Attracting Sets in Gene Regulatory Systems. In : *Proceedings SMMS 2015*, November 22-23, 2015, Chiang Mai, Thailand, pp. 135-138. DEStech Publication, Inc., Lancaster, U.S.A.

Consideration of analogies between magnetic and quantum notices for molecular network

Henryk Piech[1], *Jerzy* Jelonkiewicz[2*], *Lukasz* Laskowski[3], *Magdalena* Laskowska[4]

[123]Czestochowa University of Technology, PL-42-201 Czestochowa, Poland
[4]Institute of Nuclear Physics Polish Academy of Sciences, PL-31342 Krakow, Poland

Abstract. Magnetic properties of spin glass materials [9,13] are close to quantum interpretation in their nature description [17]. Therefore, we can look for possible kinds of analogies in process of defining theoretic and practice conventions, rules and applications of the specific characteristics in elaboration quantum calculation strategies. We have not investigated possibilities to create directly quantum calculation units and practice calculation structures like qubits, registers, gates etc. [4,18], but dealing with spin and quantum definitions and descriptions we can try to involve these notices from different domains. Such a pragmatic approach only intuitively gives chances to create the transition theory and implement it even partially. Obviously, almost all of us have heard about quantum factorization, cryptography or teleportation but it is obtained as a result of exploration casually selected quantum properties and adapting them to mathematic problems. In our approach, we carefully investigate involutions among spin and quantum nature looking at possible implementation in molecular network.

1 Introduction

One of essential problems connected with magnetic spin quantification consists in finding clear and readable approach leading to adequate physical and quantum mechanical characteristics and functions description [5,14,16]. Starting with selection of relevant structures and parameters according to quantum conversion, the mutual involvement can be exposed and practically exploited [3,9]. Widely explored experiments give us more and more information about magnetic properties of different kinds of material (spin-glass, paramagnetic, ferromagnetic etc.) molecules [2,8,13,17]. The result of experiments based on, e.g., Raman and EDX spectra [6,10,11] are agreed with theoretical analysis what confirm correctness of the models for supplementary investigations also theoretic. Carried out research can lead, for example, to conclusions about homogenous distribution of active units inside silica matrix [7,10]. On the one hand, the probabilistic aspect has to be regarded in quantum and spin magnetic processes [5,6] but on the other hand the parallel activity nature or maybe, in future, conversion is inherent feature of these conventions. In both cases (spin magnetic and quantum conventions) we should avoid quantization problem [15]. Presentation and explanation of new ideas can be graphically illustrated as selected superposition of chronology and space. We should talk about quantum calculi considering gates, Boolean function operations or computation as one with enough information on the output to deduce unambiguously from the input values [12]. Such operations on gates are essential for quantum computing also because of the quantum physics phenomena reversibility [16,20,21]. For theoretical physics quantum computing can be seen as a new challenge and also a new source of stimuli aims and paradigms how to deal with one of the most basic problems for contemporary science (physics) [17,19]. The big question is: how to resolve mysteries paradoxes and quite evident incompleteness of the most basic powerful and fascinating theory in physics, i.e. quantum theory and perhaps how to develop the main global aim of the current science, a new global theory of everything. For experimental physics especially in the area of atomic optic physics and mechanics quantum computing brings radically new challenges. We have omitted the basic ideas of classical and quantum computing including description of some experiments principles of quantum mechanics and basics of Hilbert spaces. We will start with a basic question: why to consider quantum computing? This involves introduction of basic ideas of quantum computing via a comparison of quantum and randomized computing [4,21]. Finite dimensional Hilbert spaces correspond to the example of such particle properties as spin value, polarization, position or momentum. Our proposition refers to a suggestion that outside magnetic field can play role of a controller of character logic operations, implemented by magnetic spin gates.

* Corresponding author: jerzy.jelonkiewicz@iisi.pcz.pl

2 Base formalisms representing quantum calculation in relation to spin properties presentation

Qubit is a quantum state of two level own orientations |0> and |1>, which can be presented as superposition of inner states:

$$|Y> = a|0> + b|1>$$
$$|a|^2 + |b|^2 = 1 \quad (1)$$

Qubit can be illustrated like spin in Bloch sphere (Fig.1 and 2)[4].

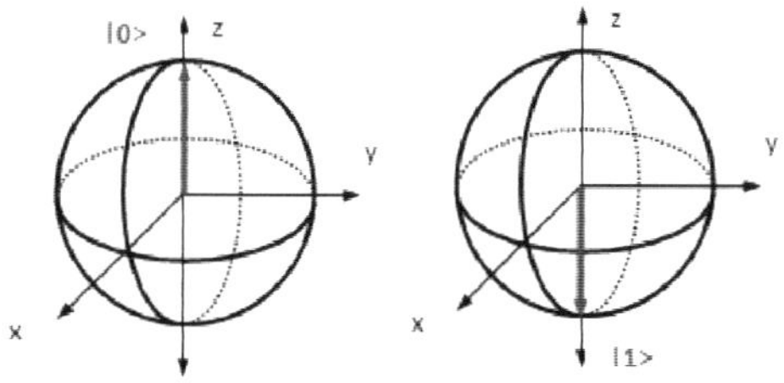

Fig. 1. Qubit illustrations: $|a|^2 = 1$ and $|b|^2 = 0$ (left) as well as $|a|^2 = 0$ and $|b|^2 = -1$ (right).

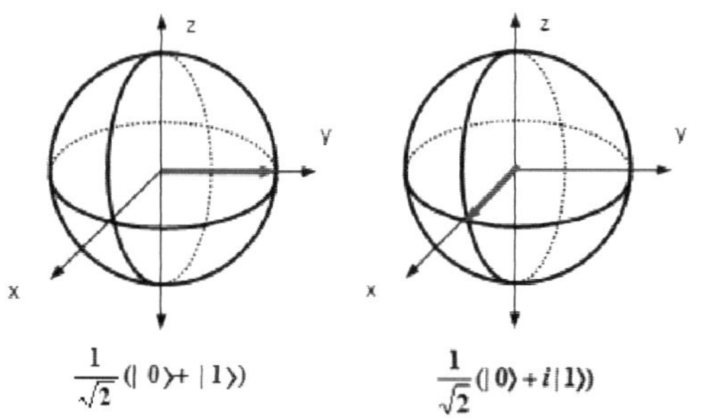

Fig. 2. Qubit illustrations: $|a|^2 = 1/2$ and $|b|^2 = 1/2$ (left)as well as $|a|^2 = 1/2$ and $|b|^2 = -1/2$ (right).

Physically we can treat qubit as a spin in magnetic field being superposition of the base states. Two level quantum state (Fig.2) has the form:

$$|\psi\rangle = cos(\frac{\theta}{2})|0\rangle + e^{i\phi} sin(\frac{\theta}{2})|1\rangle, \quad (2)$$

where e^{if} - phase coefficient, f, q - real value of phases.

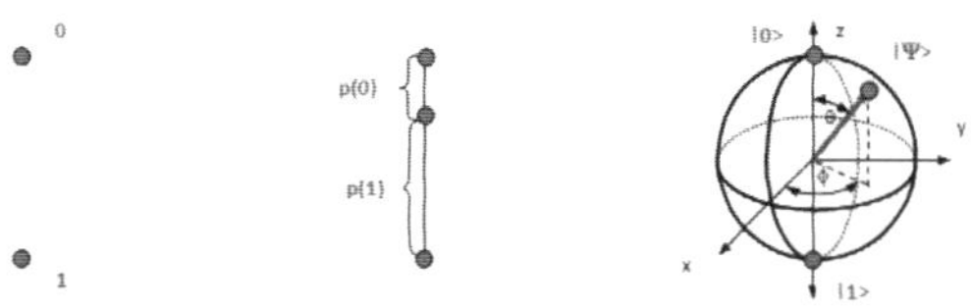

Fig.2. The classic, probabilistic (spin) and quantum states

The short explanation of spin behavior can be presented as follows: we start with chaotic spin vectors. Rotating around their own axe protons create magnetic spin moment. Total vector of magnetic spins is equal zero. Then we switch on an outside magnetic field B (Fig.3).

The part of spin vectors will allocate along field B while remained in opposite direction.

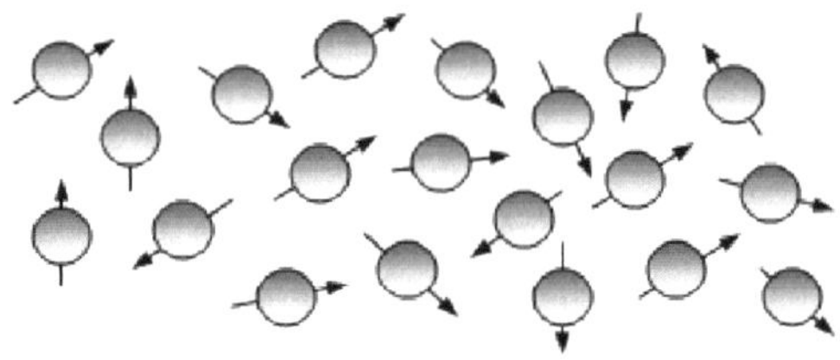

Chaotic stage

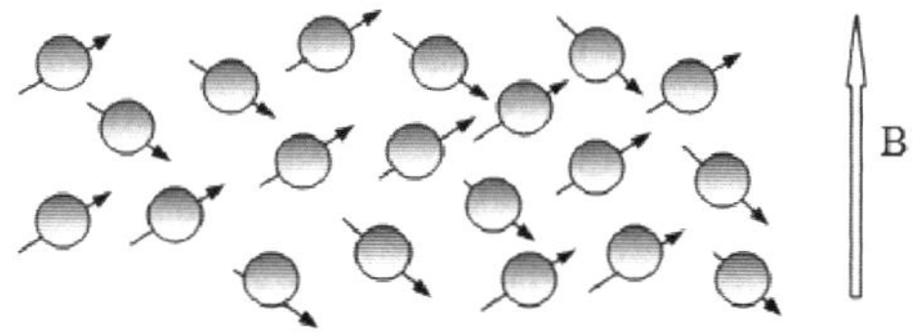

Magnetic field interference

Fig. 3. The illustration of magnetic field influence on spin vector orientation

All spinning protons are in precession, cyclic movement (Fig. 4). The speed of this motion is usually described by frequency (called Lamor frequency) which is equal to the frequency of absorbed by nucleus radiation.

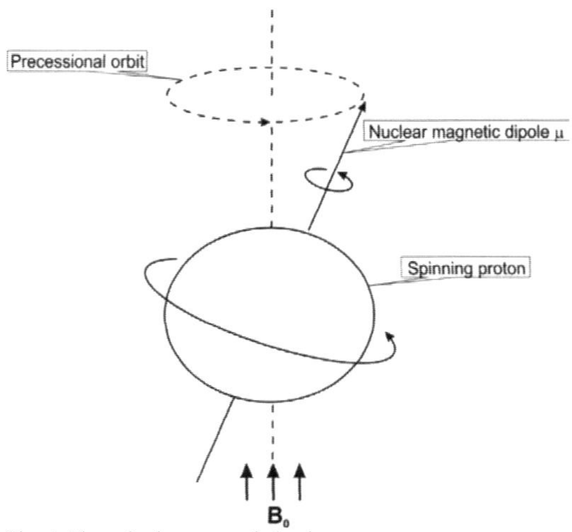

Fig. 4. The spinning proton behavior

When the field B_1 is on the magnetic vector rotates around axe x (Fig. 5a). Its new location presents Fig. 5b. After turn off radiation magnetic vector returns to its previous location (Fig.5c).

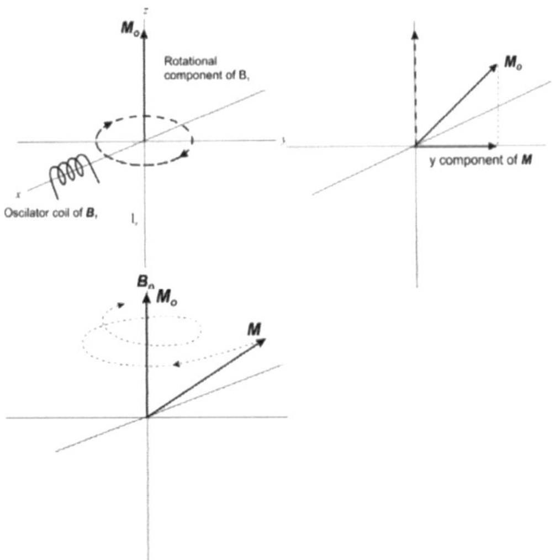

Fig 5. Magnetic vector rotations before(a), during (b) and after radiation (c).

Its probability calculation we can write: $p(1,k)=(M_0+M_0\cos(\theta+\Delta\theta))\ /(2M_0))$, where $p(1,k)$ – probability of state "1" referred to k-th molecule, θ – phase correction inferring from magnetic field impact (Fig.6). Obviously $p(0,k)=1-p(1,k)$.

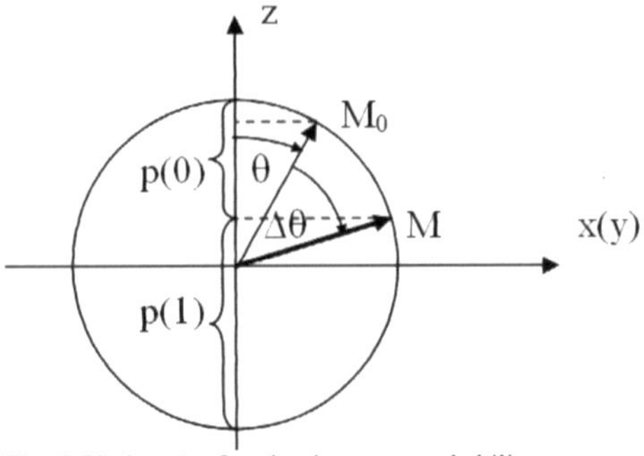

Fig. 6. Variant A of evaluation state probability

In another variant (B) we use formula: $p(1,k)=(M_0\cos(\theta)+M_0\cos(\Delta\theta))/(2M_0\cos(\theta))$ (Fig.7).

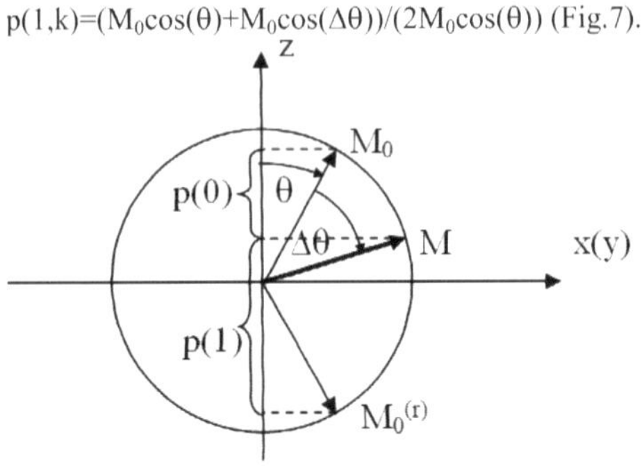

Fig. 7 Variant B of evaluation state probability

Including relatively magnetic spin changing (according to neutral location) we can propose variant C, which is represented by the formulas: $p(1,k)=(M_0-M_0\cos(\theta+\Delta\theta))\ /M_0$, $-\theta \le \Delta\theta \le 90-\theta$, or $p(1,k)=(M_0+M_0\cos(\theta+\Delta\theta))\ /M_0$, $90-\theta \le \Delta\theta \le 180-\theta$ (Fig. 8)

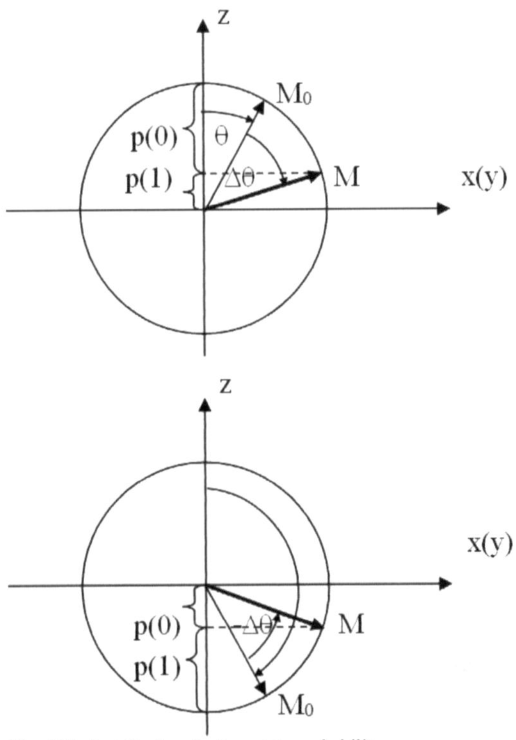

Fig. 8 Variant C of evaluation state probability

Obviously many different forms can be used for defined dependence considering probabilities spin information and magnetic state values. This information is the base of transition to quantum conversion description.

3 Generic strategy of energy spin modeling

Usually we start with Edwards-Anderson model [1]. For generalization Hamiltonian sum for all spin *S* combination is defined as:

$$H(S,J) = -\sum_{\|i-j\|=1} J(i,j) * S(i) * S(j) \tag{3}$$

where $\|*\|$ - Euclidian distance, $J=\{J(i,j)\}$- matrix of spin influences, $S(i)=\{+1,-1\}$

There are much more information available for this model but they are not important for our consideration. We supplement information about next models, for example using Sherrington-Kirkpatric model:

$$H(S,J) = -\sum_{(i,j)} J(i,j) * S(i) * S(j) - h\sum_{i} S(i) \tag{4}$$

where $S(i)=1,\dots,N$, $p(J(i,j)) \cong e^{-NJ^2/2J^2}$ h - influences of outside field energy on single spin.

The same formula is used in Ising model but *(i,j)* refers to pair of adjacent spins [2,7,8,19]. The average values of $\overline{h(i)} = 0$ as well as $\overline{h(i)h(j)} = \Delta S(i,j)$.

Magnetization degree is equal to M= $\frac{1}{N}\sum^{N} S(i)$

In p-spin model the following convention is used:

H(S,J)=

$$-\sum_{i(1),i(2),\ldots,i(r)} J(i(1),\ldots,i(p)) * S(i(1)) * S(i(2) * \ldots * S(i(r)) \quad (5)$$

where $\sum_i S(i) = N$, $p(J) \cong e^{-\frac{J^2 N^{r-1}}{r!}}$

For the above equations all spin influences are regarded. We can also consider independent pair model. To describe independent connection usually correlation function is used:

f(i,j)=<*S(i)***S(j)*> - <*S(i)*>*<*S(j)*>,

where<*S(i)*> means average over *i*-th spin states. Basing on statistic sum formalisms we get:

$$<S(0)*S(1)>= \sum_{\sigma} \frac{1}{Z} e^{-\beta * E_{01}(\sigma)} S(0) * S(1) * S(1) * S(2),$$

where Z= $\sum_{\sigma} e^{-\beta * E_{\sigma}}$ - statistical sum, σ - state, E_{σ} - state σ energy, $\beta = \frac{1}{kT}$, *k* – the number or adjacent molecules, *T* – temperature.

In the random energy model there are 2^N configurations whose energies $\{E_i\}$ are independent random variables and laid out accordingly with distribution:

$$p(E_i)=C * e^{-\frac{E_i^2}{N}}, \quad (6)$$

where $i=1,2,\ldots,2^N$, C- normalization factor.

All models refer to net structures and inner connection between them. These nets can be casual, pseudo-regular or regular [8,19]. For these type of nets main energetic and additional topologic parameters are defined. In our case parameters are not relevant but they have very important practical meaning.

4 Quantum representation and evolution

States |0> and |1> of a qubit are referred to classical states (bits). The difference between classical bits and qubits is that: classical bit can be set up only to one of the two states while a qubit can take any quantum linear superposition of |0> and |1> in uncountable states space. This means that infinite amount of information could be encoded in amplitudes of a single qubit by appropriately choosing α and β. Going to two qubits, we present this situation as a superposition of four states:

$|\Psi> = \alpha_{00}|00> + \alpha_{01}|01> + \alpha_{10}|10> + \alpha_{11}|11>$,

where $|\alpha_{00}|^2+|\alpha_{01}|^2+|\alpha_{10}|^2+|\alpha_{11}|^2 = 1$.

During the measure process of two qubits state the probability that the 1st qubit is in state *i* and 2nd in state *j* is equal to $p(i,j)=|\alpha_{i,j}|^2$. The probability that the 1st qubit is equal to 0, independently of 2nd qubit state equals $p(1^{st}\ qubit=0)=|\alpha_{00}|^2+|\alpha_{01}|^2$. The new state consists of the terms in the superposition and in the normalized form can be represented by unit vector:

$$|\phi>= \frac{\alpha_{00}|00> + \alpha_{01}|01>}{\sqrt{|\alpha_{00}|^2 + |\alpha_{01}|^2}} \quad (7)$$

The joint state of the two qubits is described in the following way [21]:

$$|\phi>=|\phi>\otimes|\phi>=(\alpha_1|0>+\beta_1|1>)\otimes(\alpha_2|0>+\beta_2|1>)=$$
$$=\alpha_1\alpha_2|00>+\alpha_1\beta_2|00>+\beta_1\alpha_2|00>+\beta_1\beta_2|00>. \quad (8)$$

The evolution of a quantum system has unitary character [4]. If we have a *s*-state quantum system then a unitary transformation is linear which can be represented by *s´s* matrix *U*. The rows and columns form orthogonal basis. U preserve inner product <*u*|*v*> between states |*u*> and |*v*>, therefore *U* preserves norms and angles. The eigenvalues of *U* are all of the form $e^{i\theta}$ [4,21]. *U* has the diagonalized form:

The evolution of a quantum system has unitary character [4]. If we have a *s*-state quantum system then a unitary transformation is linear which can be represented by *s´s* matrix *U*. The rows and columns form orthogonal basis. U preserve inner product <*u*|*v*> between states |*u*> and |*v*>, therefore *U* preserves norms and angles. The eigenvalues of *U* are all of the form $e^{i\theta}$ [4,21]. *U* has the diagonalized form:

$$\begin{pmatrix} e^{i\theta_1} & 0 & \ldots & 0 \\ 0 & \ldots & \ldots & 0 \\ \ldots & \ldots & \ldots & 0 \\ 0 & \ldots & 0 & e^{i\theta_s} \end{pmatrix} \quad (9)$$

Let us go to unitary transforms (quantum gates).

The example of gate with one qubit is Hadamar gate (a reflection around $\pi/8$ or rotation around $\pi/4$ followed by a reflection [21]) .

$$H = \frac{1}{\sqrt{2}}\begin{pmatrix} 1 & 1 \\ 1 & -1 \end{pmatrix} \quad (10)$$

H-is real and symmetric, $H^2=I$

More flexible to spin performing may occur rotation gate (according to θ):

$$U = \begin{pmatrix} \cos(\theta) & -\sin(\theta) \\ \sin(\theta) & \cos(\theta) \end{pmatrix} \quad (11)$$

Next important gates are NOT gates:

$$NOT = \begin{pmatrix} 0 & 1 \\ 1 & 0 \end{pmatrix} \quad (12)$$

and Phase Flip gates

$$PF = \begin{pmatrix} 1 & 0 \\ 0 & -1 \end{pmatrix}$$

https://doi.org/10.1051/itmconf/20181601006

The basis contains states:

$$|+>=\frac{1}{\sqrt{2}}(|0>+|->), \quad |->=\frac{1}{\sqrt{2}}(|0>-|->)$$

and we obtain: $z|+>=|->$ and $z|->=|+>$.
Moreover, we can also mention about Controlled NOT (CNOT) gate:

$$CNOT=\begin{pmatrix} 1 & 0 & 0 & 0 \\ 0 & 1 & 0 & 0 \\ 0 & 0 & 0 & 1 \\ 0 & 0 & 1 & 0 \end{pmatrix} \tag{13}$$

The first bit – control bit, the second – target bit. This gate changes qubit state if and only if the control bit is equal |1>. Therefore, we obtain:
CNOT|00> = |00>, *CNOT*|01> =|01> *CNOT*|10> = |11>, *CNOT*|11> = |10>.
There are several variants of different controlled gate - Toffoli, Fredkin, Petri etc. [4].
If we want to describe unitary transformation *k*-state system A into *l*-state system B we can use tensor product in a form A|i>⊗B|j>, where $0 \le i \le k-1$ and $0 \le j \le l-1$:

$$(\sum_{i=0}^{k-1} \alpha_i |x_i>)(\sum_{j=0}^{l-1} \beta_j |x_j>) = \sum_{i=0}^{k-1}\sum_{j=0}^{l-1} \alpha_i\beta_j |x_i>|y_j> = \sum_{i=0}^{k-1}\sum_{j=0}^{l-1} \alpha_{ij} |x_i>|y_j> \tag{14}$$

So called untangled system and lexicographical matrix basis for two bits structures is the following:

$$\begin{pmatrix} a_{11}B & a_{12}B & a_{1n}B \\ a_{21}B & a_{22}B & a_{2n}B \\ & & \\ a_{m1}B & a_{m2}B & a_{mn}B \end{pmatrix} \tag{15}$$

where $B=\{b_{i,j}\}$, $1 \le i \le k$ and $1 \le j \le l$.

5 The effort of gates creation for spin nature elements

In this section we do not describe synthesis process of microporos or mesoporous materials but only use them basing on their microstructures and features. The microstructure of mesoporous silica (SBA-15) in simplified form is presented in Fig. 9.

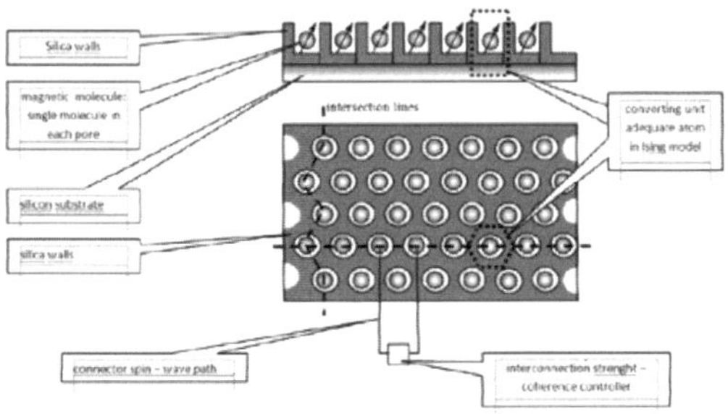

Fig.9. The structure model of mesoporous silica

Generally, magnetic molecules are mutually energetically isolated (SBA-15) or partially connected (ferromagnetic, paramagnetic materials), creating network with casual structure. Theoretically, it is possible to build energetic interconnection between molecules (coherent controller) based on spin-FET (field effect transistor design [1]) and create a molecular network. In spite of this, our proposition leads to similar solution. Basing on magnetic properties, we propose to apply influences outside energetic field on spin state and use these possibilities to encode probabilistic binary state. Consequently, we suggest the implementation of fundamental logic conversion performed by gates with a help of outside magnetic field controller. Let us consider the simplest logic operations (conjunction, disjunction). The idea of our proposition can be presented as follows: in initial state spin moment *M* is not strictly defined, so we assume that its position (q) is as in Fig.10.

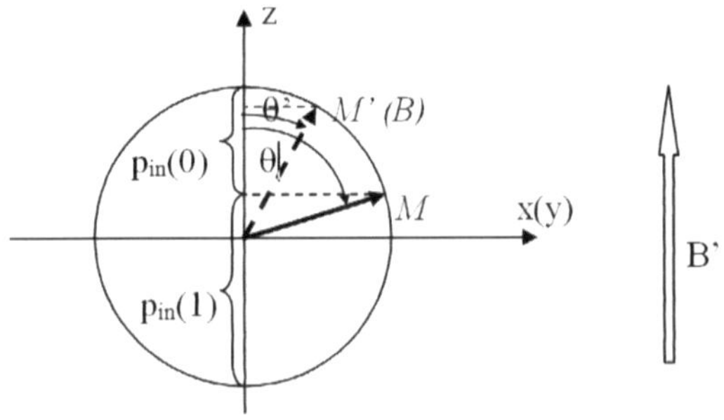

Fig. 10 The effect of turning on outside magnetic field B' evoking phase correction θ'< θ and adequate probability correction $p'(1) > p_{in}(1)$.

Turning on outside magnetic field B' we cause the change of value and phase of magnetic spin vector. After scaling procedure we will get the vector *M'* (dashed arrow in Fig.10). The vector *M'* refers to bit "1" (or to qubit |1>) with probability $p'(1) > p_{in}(1)$. The angle θ depends on strength of magnetization. In the similar way we form information that refers to bit "0" (or adequate qubit) what is presented in Fig. 11. On this stage of consideration the value of spin vector do not play any relevant role.

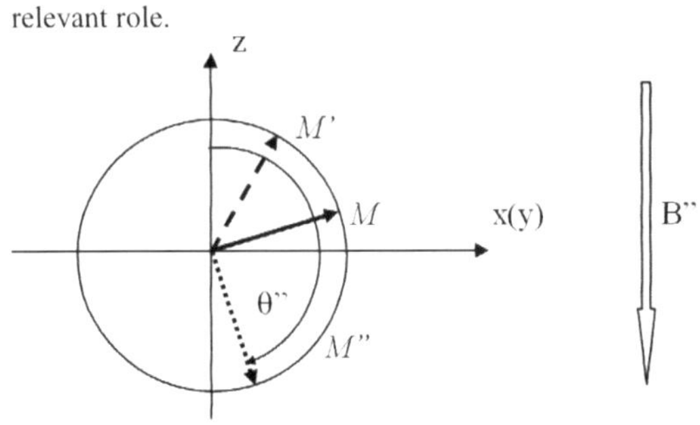

Fig. 11 The effect of turning on outside magnetic field B' evoking phase correction θ''> θ and adequate probability correction $p''(0) > p_{in}(0)$.

Thus a spin vector *M''* appears (doted arrow) that is adequate to "0" bit information. We assume, that absolute values of magnetization strength are in relation |B''|>|B'|. Therefore, the phases are in relation |θ'|> |90-θ''|. After turn off the field previous magnetic characteristics are preserved for relaxation time [13].

The next stage of gates building is connected with energetic interconnections in stochastic (or statistic) way. It is presented in a diagram as a general effect of mutual molecules influences to show the idea of gates creation. Firstly, this refers to conjunction gate functioning (Fig.12) in which main controller assumption is |θ'|> |90-θ''|.

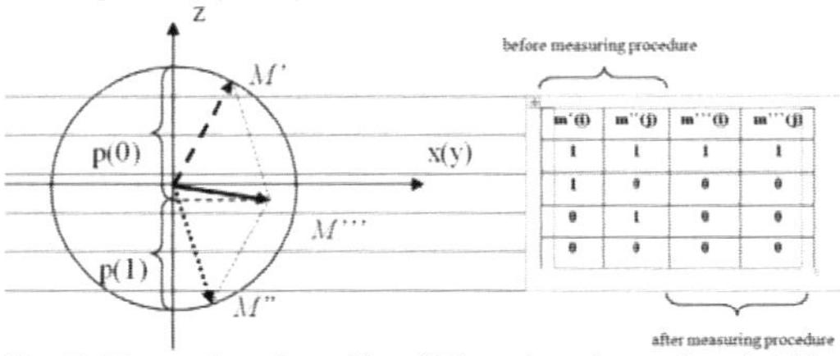

Fig.12 The conjunction effect (CC gate) on base of probability dependency: (p(0)>p(1))~ bit "0" (qubit |0>), (p(1)>p(0))~ bit "1" (qubit |1>).

Secondly, we can refer to disjunction gates. In this case we assume that |θ'|< |90-θ''|.
The adequate situation is presented in Fig.13.

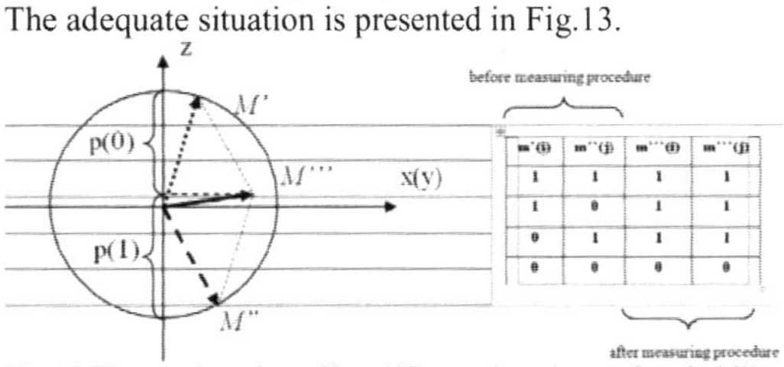

Fig.13 The conjunction effect (CD gate) on base of probability dependency: (p(0)>p(1))~ bit "0" (qubit |0>), (p(1)>p(0))~ bit "1" (qubit |1>).

To negate initial (or given) state we use the opposite directed energetic fields (Fig.14). The strength of its influences on spins should follow the condition θ'''>90-θ'.

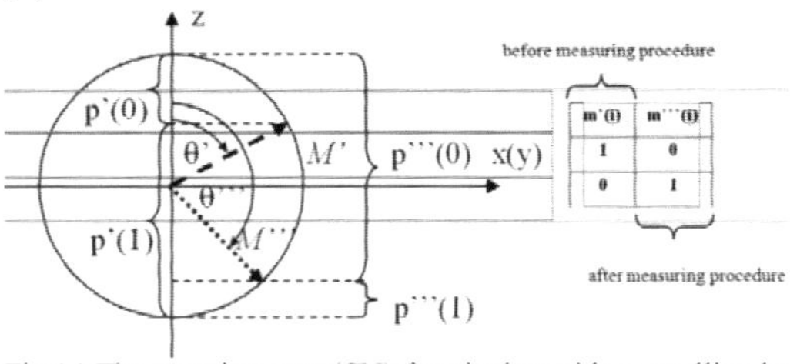

Fig.14 The negation gate (CN) functioning with controlling by outside magnetic field

The graphic presentation of control spin gates can be similar as quantum gates (Fig. 15).

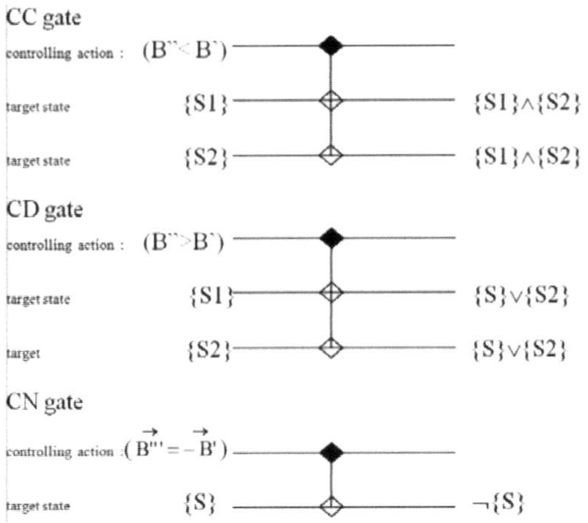

Fig.15 The provisory of spin gate illustration.

Naturally, we assume that initial state of spin magnetic parameter is known or can be read or measured. In similar way we can project different kinds of gates.

6 Conclusions

We hope that finding analogies between quantum and spin nature of molecules exposed to magnetic field can shorten the distance between theory and practice. The proposition of using outside magnetic field to control character of logic conversion in stochastic approach offers a chance to quantum calculation possibilities. Technically this is much more difficult because of many implementation limits and constrains connected with spin nature character [1,6]. Selection of the materials, considered parameters and also sources of outside energy creates quite a large research space. Perhaps, proposed approach of creation controlled elementary logic gates guaranties combining the bests features of traditional, quantum and nanostructure convention leading to molecular network.

References

1. Agnihotri P., Bandyopadhaya S.: Analysis of the two dimensional data-das spin field effect transistor, *Physica E: Low-dimensional Systems and Nanostructures* **42(5)**,1736-1740 (2010)
2. Albayark E., Yigit A., and Akkaya S.: The antiferromagnetic Ising model for a bilayer Bethe lattice, *Journal of Magnetism and Magnetic Materials* **320** 2241-2248, (2008)
3. Bonesteel N.E., Stepanenko D., and Di Vincenzo D.P.: Anisotropic spin exchange in pulsed quantum gates, *qu-phys* **0002092** (2000)
4. Brylinski L.L., and Brylinski R.: Universal quantum gates, qu-phys **0108062** (2001)
5. Childs A. and Chuang I.: Universal quantum computation with two-level trapped joins *qu-phys* **0008065** (2000)
6. Cornia A., Mannini M., Sainctavit P., Sessoli R.: Chemical strategies and characterization tools for

the organization of single molecule magnets on surfaces, *Chem. Soc. Rev.* **40**, 3076-3091 (2011)
7. Dorogovtsev S., Goltsev A.V., and Mendes J.F.F.: Ising model on networks with an arbitrary distribution of connections, *Physical Review* **E 66:016104** (2002)
8. Herrero C.P.: Antiferromagnetic Ising model in small-world networks, *Physical Review* **E 77:041102** (2008)
9. Lacroix C.: Magnetic properties of strongly frustrated and correlated systems, *Physica B: Physic and Condensed Matter* **404** 3038-3041 (2009)
10. Laskowski L., Laskowska M.: Functionalization of SBA-15 mesoporous silica by cu- phosphonate units: Probing of synthesis route. *Journal of Solid State Chemistry*, **220**, 221-226 (2014)
11. Laskowski L., Laskowska M., , Balanda M., Fitta M., Kwiatkowska J., Dzilinski K., Kaczmarska A: Mesoporouse silica SBA-15 functionalized by Nikel-phosphonic units: Raman and magnetic analysis. *Microporous and Mesoporous Materials* **200**, 253-259 (2014)
12. Malarz K., Antosiewicz W., Karpińska J., Kułakowski K.: Avalanches in complex spin networks, *Physica A* **373**, 785-795 (2007)
13. Parisi G.: Some consideration of finite dimensional spin glasses, *Journal of Physics* **A 41:324002** (2008)
14. Piech H., Siedlecka-Lamch O.: Parallel Quantum Algorithm for finding the consistency of Saaty's matrices, *Lecture Notes in Computer Science* **7203** 102-111 (2012)
15. Schlieman J., and Loss D.: Entanglement and quantum gate operations with spin-qubits in quantum dots, cond- mat 0110150 (2001)
16. Solano E., Santos M.F., and Milman P.: Quantum phase gate with selective interaction, *qu-phys* **0102052** (2001)
17. Stein D.L.: Spin glasses: Still complex after all these years? *Lecture Notes in Physics* **633**, 349-361 (2003)
18. Thorwat M., and Haenggi P.: Decoherence and dissipation during quantum XOR gate operation, *cond-mat* **0104513** (2001)
19. Timonin P.N.: Thermodynamics of strongly frustrated magnet in a field: Ising antiferromagnetic on triangular lattice, *arXiv*:**1005.2650v1** (2010)
20. Vidal G., Masanes L., and Cirac J.I.: Storing quantum dynamics in quantum states: syochastic programmable gate for u(1)operations, *qu-phys* **0102037** (2001)
21. Zhou X., Leung D., and Chuang I.: Methodology of quantum logic gate construction, *qu-phys* **0002039** (2000)

CICT: New Eyes on Computational Competence in Computational Science

Rodolfo A. Fiorini*

Politecnico di Milano University, 20133 Milano, Italy

Abstract. Science does not exists to enlighten people's minds only. It mainly exists to show the educated way from quanta to qualia. And that way starts from computational competence. In previous papers published elsewhere, we have already shown that traditional Q Arithmetic can be regarded as a highly sophisticated open logic, powerful and flexible bidirectional formal language of languages, according to "Computational Information Conservation Theory" (*CICT*) new perspective. This new awareness can offer competitive approach to guide more effective and convenient algorithm development and application to arbitrary multiscale (AMS) biomedical system modeling and simulation. An articulated example on function computational modelling is presented and compared to standard, well-known and traditional approach. Results are critically discussed.

1 Introduction

In the past decades, we learned how traditional human-made system can be quite fragile [1] to unexpected perturbation, because statistics by itself can fool you, unfortunately. In fact, they were developed and used under the implicit assumption of classic model completeness. Traditional models of Systems Theory were intended to completely represent aspects of phenomena and processes, such as the motion of a pendulum or the operation of an amplifier. They concern the phenomena in their temporal and spatial completeness [2]. The current successful extraction of scientific knowledge from experimentation is based on computer modeling and simulation mostly. Classically, the possible incompleteness in the modeling is assumed as having a provisional or practical nature as being still under study and because there is no theoretical reason why the modeling can not be complete.

Concepts and approaches regarding contexts and processes for which systems modeling can not be conceptually exhaustive have been already introduced in the literature [3]. We recall, first of all, fuzzy sets and fuzzy logic [4,5] for which, however, completeness has merely a probabilistic nature. It is matter of classical, computable probability. Nevertheless, there are phenomena, which must be modeled regarding their coherence being a crucial systemic theme with respect to their completeness or comprehensiveness as considered by Logical Openness [6] and computational information conservation theory (*CICT*) [7] for arbitrary multiscale (AMS) [8] system issues. Therefore, we must consider contexts and processes for which modelling through the use of systems is incomplete since related to only some properties, as well as those for which such modeling is theoretically incomplete as in the case of processes of emergence and for approaches considered by the Second Systemics [9]. As an example, in advanced biomedical applications [10, 11], sociology and social cybernetics [12], to ascertain complex causality reliably is always problematic, because the usual external observations always reveal superficial reasons only; they cannot reveal deep, concealed reasons [11, 13]. Furthermore, it would be interesting to consider how theoretical incompleteness, incomplete modelling, i.e., not exhausted by using individual models, of processes and phenomena should be explored as a conceptual coexistence of different approaches not so much with the purpose of exhausting, but to conceptually represent the structural dynamics of becoming [14].

At the 15th IEEE International Conference on Cognitive Informatics & Cognitive Computing, the author presented a paper introducing neuromorphic ALS (anticipatory learning system) [15], where he suggested that to achieve reliable system intelligence outstanding results, current computational system modeling and simulation has to face and to overcome two orders of issues at least, as soon as possible:

1- To minimize the traditional limitation of current digital computational resources that are unable to capture and to manage even the full information content of a single Rational Number Q exactly, leading to information dissipation and opacity.

2- To develop stronger, more effective and reliable correlates by correct arbitrary multiscale (AMS) [8] modeling approach to complex system.

*Corresponding author: rodolfo.fiorini@polimi.it

To address effectively the two above modeling issues, we need a transdisciplinary approach first, to become deeply aware that human beings and most current advanced computational and measurement systems formal representations have quite a limited capacity to capture and to extract reliable information from noisy sources or generators traditionally. This is mostly due to the representation ambiguity at the core of traditional and current computational mathematics [16].

2 Symbolic vs. Operational Number Representation

In computational mathematics, computer algebra (CAL), also called symbolic computation or algebraic computation, is a scientific area that refers to the study and development of algorithms and software for manipulating mathematical expressions and other mathematical objects. Algebra started with methods of solving problems in arithmetic. The numerical symbols consisted probably of strokes or notches cut in wood or stone, and intelligible alike to all nations. For example, one notch in a bone represented one animal, or person, or anything else. The Egyptians paid attention to geometry and numbers, and the Phoenicians to practical arithmetic, book-keeping, navigation, and land-surveying. Numerical notation's distinctive feature, i.e. symbols having local as well as intrinsic values (arithmetic), implies a state of human civilization at the period of its invention. At the beginning of CAL, circa 1970, when the long-known algorithms were first put on computers, they turned out to be highly inefficient [17]. Therefore, a large part of the work of the researchers in the field consisted in revisiting classical algebra in order to make it effective and to discover efficient algorithms to implement this effectiveness. A typical example of this kind of work is the computation of polynomial greatest common divisors (GCD), which is required to simplify fractions. Surprisingly, the classical Euclid's algorithm turned out to be inefficient for polynomials over infinite fields, and thus new algorithms needed to be developed. The same was also true for the classical algorithms from linear algebra.

Although, properly speaking, CAL should be a subfield of scientific computing, they are generally considered as distinct fields because scientific computing is usually based on numerical computation with approximate floating point numbers, while symbolic computation emphasizes exact symbolic computation with expressions containing variables that have no given value and are manipulated as symbols, hence the name "symbolic computation" (SC). SC has also been referred to, in the past, as "symbolic manipulation", "algebraic manipulation", "symbolic processing", "symbolic mathematics", or "symbolic algebra", but these terms, which also refer to non-computational manipulation, are no more in use for referring to computer algebra.

CAL is widely used to experiment in mathematics and to design the formulas that are used in numerical programs. It is also used for complete scientific computations, when purely numerical methods fail, like in public key cryptography or for some non-linear problems. As numerical software are highly efficient for approximate numerical computation, it is common, in CAL, to emphasize on exact computation with exactly represented data. Such an exact representation implies that, even when the size of the output is small, the intermediate symbolic data generated during a computation may grow in an unpredictable way. This behavior is called "expression swell." To obviate this problem, various methods are used in the representation of the data, as well as in the algorithms that manipulate them.

The usual numbers systems used in numerical computation are either the floating point numbers and the integers of a fixed bounded size, that are improperly called integers by most programming languages. None is convenient for CAL, because of the expression swell. Therefore, the basic numbers used in CAL are the integers of the mathematicians, commonly represented by an unbounded signed sequence of digits in some base of numeration, usually the largest base allowed by the machine word. These integers allow to define the rational numbers $\boldsymbol{Q}$, which are irreducible fractions of two integers.

Computer arithmetic (CAR) is a branch of computer engineering that deals with methods of representing integers and real values (e.g., fixed- and floating-point numbers) in digital systems and efficient algorithms for manipulating such numbers by means of hardware circuits or software routines. Programming an efficient implementation of the arithmetic operations to obtain a reliable numeric representation is a hard task usually. In fact, a current intriguing point is that, although there are multiple models for the integer numbers, they all will agree on the definition of computable functions. However, current real number computation does not have these properties. Traditional scientific computation, based on classic CAR, uses specified fixed-length finite representations (related to scientific notation) of real numbers, and so can achieve only limited precision, can make errors in comparisons, and can even be unstable over rounds of conversion to and from corresponding decimal representation. Amazingly, whether an extended Turing machine model or a real-number computation model is appropriate for scientific computation is still an open topic of discussion. Therefore, most free computer symbolic algebra systems and some commercial ones, like Maple (software), use the GMP library, which is thus a de facto standard [18].

Number theory or, in older usage, arithmetic is a branch of pure mathematics devoted primarily to the study of the integers. It is sometimes called "The Queen of Mathematics" because of its foundational place in the discipline. In fact, by arithmetic, Plato meant, not arithmetic in our sense, but the science which considers numbers in themselves, in other words, what we mean by the Theory of Numbers. The word "arithmetic" is used by the general public to mean "elementary calculations"; it has also acquired other meanings in mathematical logic, as in Peano arithmetic, and computer science, as in floating point arithmetic). The use of the term arithmetic for number theory regained some ground in the second half of the 20th century, arguably in part due to French influence. In particular, "arithmetical" is preferred as an adjective to "number-theoretic." Arithmetic combinatorics (AC) is a field in the intersection of number theory, combinatorics, ergodic theory and harmonic analysis. Arithmetic combinatorics is about combinatorial computation and estimates associated with arithmetic operations (addition, subtraction, multiplication, and division). "Additive combinatorics" is the special case when only the operations of addition and subtraction are involved [19]. Number theorists study prime numbers as well as the properties of objects made out of integers (e.g., rational numbers) or defined as generalizations of the integers (e.g., algebraic integers). Questions in number theory are often best understood through the study of analytical objects (e.g., the Riemann zeta function) that encode properties of the integers, primes or other number-theoretic objects in some fashion (analytic number theory). One may also study real numbers in relation to rational numbers, e.g., as approximated by the latter (Diophantine approximation). In numeric representation of rational number $\boldsymbol{Q}$, rational proper quotient is represented by infinite repetition of a basic digit cycle, called "reptend" (the repeating decimal part) [20].

According to *CICT* new perspective, the first repetition of basic digit cycle corresponds to the first full scale interval where number information can be conserved completely, and *CICT* calls it "representation fundamental domain" (RFD) [21]. In this case the new computational representation of rational number $\boldsymbol{Q}$, is called "OpeRational" (OR) Representation, just to remember that *CICT* is able to conserve Rational Number $\boldsymbol{Q}$ full information content, much better than past computational approaches [22]. It is even possible to show a new point of view for number processing, where the complex upper half-plane can naturally emerge by the interplay of the coupled inversion of two counter oriented upper quarter planes. That representation can be even read as the self-reflexion of a reciprocal conformal relation of an "Outer Symbolic Representation" (OSR) to its corresponding reflected fundamental "Inner OpeRational Representation" (IOR) and vice-versa, where D in $\boldsymbol{Z}$, $A = 1/D$ and $D = 1/A$, and $DA = AD = \overline{0}1.\overline{0}$, for positive oriented upper quarter-space (right upper quarter plane) and $DA = AD = -\overline{0}1.\overline{0}$, for negative oriented upper quarter-space (left upper quarter plane) [21]. The relationship of these two counter-oriented upper quarter-spaces can be thought as a reciprocally projective conformal correspondence to be the first operative example in literature for explicit self-reflexive oriented numeric representation space [23]. This space can define a powerful reference framework to develop OR numerical applications for maximum information conservation (i.e. minimal information entropy generation and information dissipation) [24]. For the sake of simplicity, we consider main properties of one oriented upper quarter-space only, because the second one is just its reflection with respect a vertical axis, so LTR (Left-To-Right) representation turns into RTL (Right-To-Left) representation and vice-versa. A direct representation turns into additive complemented representation and vice-versa, when needed. In virtue of this relationship, the same information content can get two different numerical representations in direct space (DS) and reciprocal space (RS), across the outer universe-inner universe (OU-IU) boundary (Fig.1) which can resonate to each other freely, at the same time, with respect to an inversive reference boundary element (i.e., centered unitary segment in 1-d, unitary circumference in 2-d, unitary sphere in 3-d, unitary ipersphere in 4-d, etc.) between an nth-dimensional "Outer Space" representation and a correspondent "Inner Space" representation: an nth-dimensional OSR to its resonating IOR and vice-versa.

Therefore, the well-posed arithmetic structural description is self-defined by its functional closures. Naturally, we have two different types of closures: the linear and the exponential one [25]. This new awareness opens the way to new computational competence in computational science.

3 New Eyes on Computational Science

In the previous section we saw that current computer computation must be either symbolic or approximate. Nevertheless it can be shown that computer computation can use either an "approximated approximation", as in the past approaches, or "exact approximation" representation system, as proposed by *CICT* [21]. To achieve exact approximation computational number representation, logic must be described in terms of "closure spaces." From traditional information modeling point of view, the main focus is on the DS representation only (Euclidean space). Nevertheless, according to *CICT* ODR approach [26], to grasp the full information content of phenomenic reality, DS is just half of the OU human representation (sharable representation) and its co-direct space (CS) is the other half, the DS natural closure. Coupled to the OU is the IU human representation (subjective representation), composed by the RS and its natural closure, the reciprocal co-space (RC), or the DS dual. DS and CS are the coupled, complementary, asymptotic components of the fundamental, irreducible dichotomy of human OU representation (Fig. 1) [16, 27].

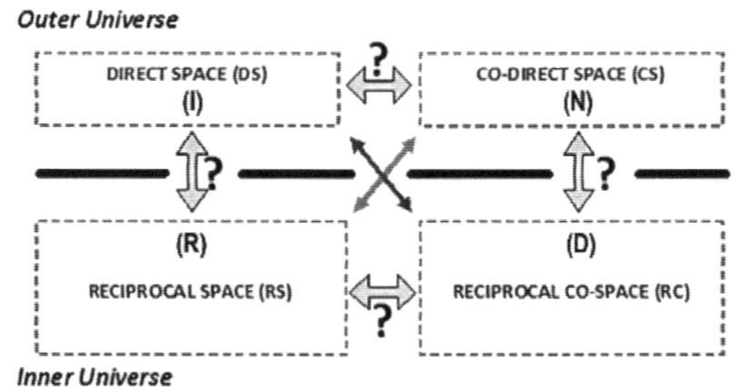

Fig. 1. The fundamental blocks of the *CICT* Outer Universe-Inner Universe Boundary approach (IOU Diagram) for full computational information conservation [16, 27] (see text).

This fundamental representation is based on two root conceptual components: unfolded information (linear sharable information that can be communicated in a formal way by media) and folded information (complex subjective information that cannot be communicated by any traditional media) [28]. According to *CICT*, this is the minimum framework to capture and to conserve information efficiently. Therefore, the best traditional DS representation can capture at maximum 25% only of the full information available theoretically. Obviously, considering these additional relationships increases the sophistication of the representation system.

On the other side, we should consider the relation between our continuous perception of the OU and the discrete representation of it. Human perception and representation of our universe can be mapped as in Fig. 1, where the encoding process is carried out by human affectors (our biological sensors) and the decoding process is done by human effectors (our biological actuators). Human biological transducers, by which we acquire information on the outer world interacting with it, are intrinsically discrete. This means that our perception of continuous shapes is just an illusion created by our brain. From this ground we can infer that an illusion of continuity can be achieved with discrete supports, without even noticing any difference, maintaining thus a maximum representation coverage property. The discrete approach reveals to be, in this sense, highly convenient because it strongly decreases the computational cost and the complexity of the system for representation modeling. According to *CICT*, in arithmetic the structure of closure spaces (across the Outer-Inner Universe boundary) is self-defined by Natural Numbers Reciprocal Space (RS) representation (Fig. 1). Furthermore, the four fundamental components of the IOU diagram (DS, CS, RS, DS) can be mapped directly to the four fundamental transformations of the Piaget-Klein four-group (I, N, R, D) to achieve predicative and reasoning competence [27].

4 Modeling OpeRational Example

The rich operative scenario offered by combinatorial modular group theory is full of articulated solutions to information processing problems. *CICT* framework is quite flexible and can be used under two major operational representation schemes: formal recurrence sequence (FRS) and formal power series (FPS) presentation, respectively. *CICT* can be presented and used in term of recurrence relation. In this case, rational geometric series can be seen as simple recurrence sequences in a wider recurrence operative framework where all algebraic recurrence sequences of any countable higher order include all the lower order ones and they can be optimally mapped to rational number system Q operational representations (OR) and generating functions (OECS, Optimized Exponential Cyclic Sequences [16]). Basic *CICT* result can be even presented in term of classic either monopolar or bipolar power series, or in general by formal power series, to show the close relationships to classic and modern control theory approaches for causal continuous-time and discrete-time linear systems [21].

Usually, from a system modeling computational perspective, all approaches that use a top-down (TD) AMS point of view allow for starting from an exact global solution of analytic solution families, which offers a shallow local solution computational precision to real specific needs (in other words, from global to local perspective overall system information is not conserved, as misplaced precision leads to information dissipation, ambiguity, confusion and system decoherence [8], [29]). This is main due to the limitations of traditional numeric computational resources as presented in previous section 2. For this reason, usually further analysis and validation (by probabilistic and stochastic methods) is always necessary to work out a localized computational solution of any practical value, in real application. A local discrete solution is achieved and computationally approximated as the last step in their line of model reasoning, that started from an overall continuous system approach (from continuum to discreteness). In this way, statistical and applied probabilistic theory became the core of classic scientific knowledge in engineering applications at system macroscale level. They were applied to all branches of human knowledge under the "continuum hypothesis" (CH) assumption, reaching highly sophistication development, and a worldwide audience. Many classic "Science 1.0" researchers and scientists up to scientific journals assume it is the ultimate language of science and it is the traditional instrument of risk-taking. Unfortunately, the "probabilistic veil" can be very opaque computationally, in a continuum-discrete arbitrary multi-scale (AMS) environment [1].

As a matter of fact, under the discreteness hypothesis (DH) with a discrete support of arbitrary precision, increasing the representation precision, we can arrive to an illusion of function continuity, offering quite interesting computational savings over the classic CH approach. Traditionally, the discreteness approach, developed under the DH assumption, in specific scientific disciplines, has been considered in peculiar application areas only. It has been further slowly developed by a few specialists and less understood by a wider audience. Unfortunately, over the centuries, the above two large scientific research areas (CH based and DH based) have followed separate mathematical development paths with no or quite little, inconsistent synergic coupling. That is the main reason why quantum field theory (QFT) modeling applications are still mostly overlooked by traditional scientific and engineering researchers for AMS system modeling, from system nano-microscale to macroscale [30].

Taking advantage of fundamental power series and polynomial system properties awareness, as achieved by *CICT* [26], here, we like to show how the concept of numerical precision can be used to control the transition from discrete to continuous in function modeling. Traditionally, in computational science, the precision of a numerical quantity is a measure of the detail in which the quantity is expressed. This is usually measured in bits, but sometimes in decimal digits. It is related to precision in mathematics, which describes the number of digits that are used to express a practical value. *CICT* uses the precision of numerical quantity to model the discreteness to continuum function transition quite easily. Our computational example has been programmed in MATLAB® language. In this way, it is possible to verify and validate through numerical computation the final results.

4.1. Example Definition

For the purpose of generating a discrete power series related example, we define the required specification as an exercise to be solved and then present the solution. Let us define the function presentation interval in the [0; 10] span and a LTR vector with a linear, fixed step Δx. This will be the axis of abscissae of the presentation window. Then, we define the values of the axis of the ordinates as the fraction of this LTR vector and a second vector (its reflection with respect to the vertical axis) in the RTL direction.
For example, if we fix the step $\Delta x = 1.0$, the representation window step is equal to an arbitrary unitary step value. We define the two vectors as follows:

$$LTR = \{0,1,2,3,4,5,6,7,8,9,10\}$$
$$RTL = \{10,9,8,7,6,5,4,3,2,1,0\} \quad (01)$$

then we compute the ration of their components with the same index respectively. We obtain:

$$values = \frac{LTR}{RTL} = \left\{\frac{0}{10},\frac{1}{9},\frac{2}{8},\frac{3}{7},\frac{4}{6},\frac{5}{5},\frac{6}{4},\frac{7}{3},\frac{8}{2},\frac{9}{1},\frac{10}{0},\right\} = \left\{0,\frac{1}{9},\frac{1}{4},\frac{3}{7},\frac{2}{3},1,\frac{3}{2},\frac{7}{3},4,9,\infty\right\}$$

This are the starting function model (FM) values of our example. Now, we formulate four operative requests as follow.

Requests:

A. Find a parameter "a" to describe this envelope in the form a^x and compare it to the previously generated values. Plot the results.

B. Increase the number of steps (precision) in the same abscissae interval [0; 10] until you have the appearance of a continuous envelope in a plot.

C. Transform the steps and the axes to make the plot of this function appear linear.

D. Adjust the precision steps to obtain a distribution of the values in the axis of the ordinates that resembles the appearance of a continuous envelope in the entire abscissae interval [0; 10], with a convenient amount of computation for each tract on the axis of the abscissae.

4.2. Results and discussion

First, it is useful to observe the shape of the discrete envelope of this FM. It looks like an exponential-like-shaped function (Fig. 2).

4.2.1 Point A Result

At starting, we keep the step equal to the initial FM, step $\Delta x = 1$, and plot the result of the values, with respect to the axis of abscissae previously defined. Observing Fig. 2, first we should notice that the tenth value is not showed. This is due to the fact that the infinite value obtained in the tenth position is too far away from the previous ninth value: plotting it, would stop us from seeing the real envelope clearly.

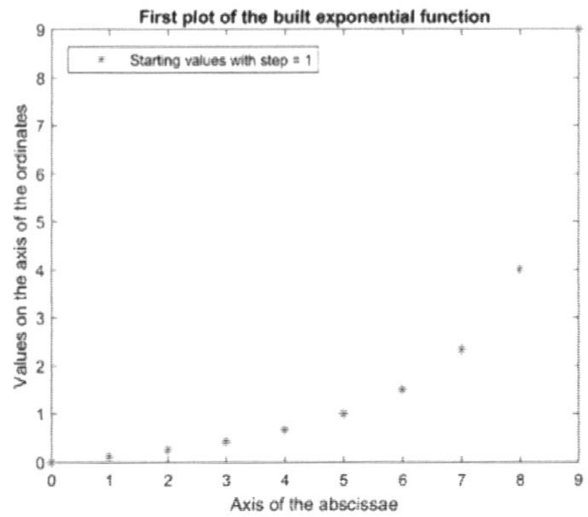

Fig. 2. Plot of the discrete exponential-like-shaped FM with unitary step.

Considering, anyway, the fact that the infinite value is present, we can identify a numerical symmetry at point (5, 1). Translating ideally the origin at this point could simplify the resolution of the Point A request. In fact, we know that the required model has the shape of an exponential function a^x and we know that it is number symmetrical at point (5, 1). Thus, once translated to the origin, we can express this numeric symmetry as $(x+5) = a^x(x-5)$. This brings us to conclude that our parameter a is equal to:

$$a = \left(\frac{x+5}{x-5}\right)^{-x} \quad (02)$$

It is a variable parameter function of x. In order to verify this relation, we can plot the values obtained with the model against the LTR/RTL values of our starting envelope of Fig. 2. The result is shown in Fig. 3 and it is clear that the two sets of points overlap, thus the model appears accurate. Obviously, the FM is fed with an input interval [-5; 5] respecting the assumed position of the origin, and then translated into the initial [0; 10] interval for the comparison.

For a better understanding of the FM properties, we can compare it with respect to a power of ten envelope and to a discrete natural exponential shape. In Fig. 4, initial FM, with step $\Delta x = 1$, is compared to the discrete natural exponential function and the 10^n behavior, evaluated only in the considered LTR points.

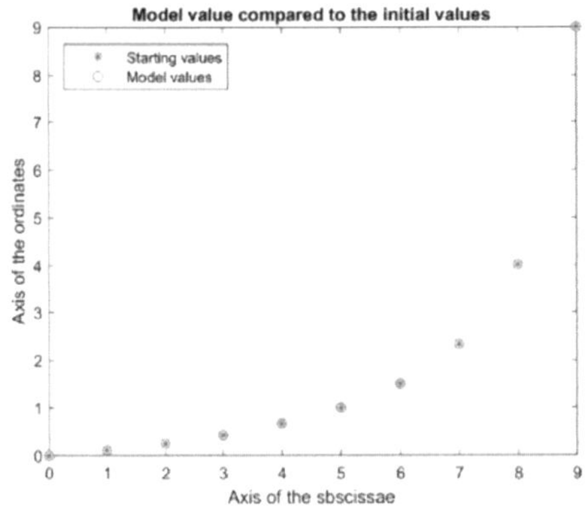

Fig. 3. Comparison between the initial FM model values and the model with the identification of parameter "a" as from eq.(02), with unitary step (see text).

Obviously, the 10^n values diverge larger from FM values, especially toward the *LTR* end of the representation window. The exponential values, instead, are closer to initial FM. Therefore, we can say that the initial FM is a discrete representation of a natural exponential function. The advantages of this representation are many: with a discrete approach it is possible to decide the level of precision at which we want to stop, with respect to the used support. The right level of precision with respect to the used representation support resolution allows to obtain the illusion of continuity in the final FM representation, with a reduced computational effort in comparison to the usual continuous approach.

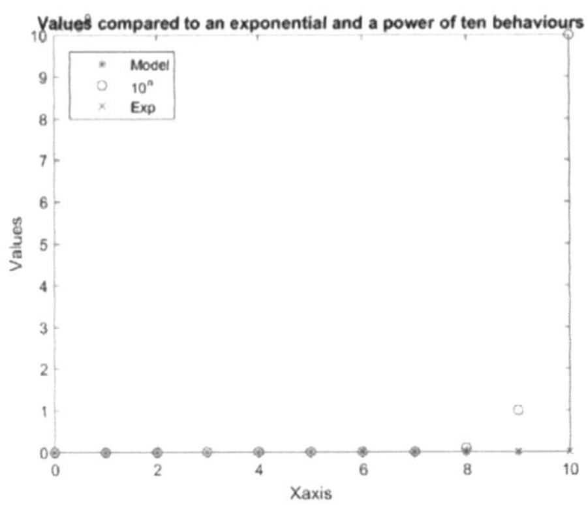

Fig. 4. Comparison of the initial FM (Model) generated values with two well-known discrete function behaviors (see text).

Furthermore, with the discrete power series approach, at each point it is possible to compute a complement to a reference value (see Section 3 for CS and RC issues), while a traditional continuous approach does not contemplate this opportunity (it offers an approximated approximation only).

4.2.2 Point B Result

For dealing with the request of obtaining the illusion of continuity, first it is possible to create a function that can be used to obtain values of the FM and the starting system as the step changes. This function, called *Ex1*, accepts as input the value of the step and returns three discrete arrays: one describing the common axis of the abscissae and the other two describing the axis of ordinates. In this way, the differences at arbitrary step values can be computed. We use *Ex1* with different values from $\Delta x=10^{-1}$ to $\Delta x=10^{-5}$ noticing that every time the step is decreased, the illusion of continuity increases and widens from the value 0.0 towards the right end of the considered representation window. The points are closer, on the axis of the ordinates, at the left beginning of the representation window and become sparse at the right end. Fig. 5 shows this behavior with step Δx = 0.001 = $10=^{-3}$. Looking at the plot, the illusion of continuity with this step is achieved until the last interval [9; 10]. The value x = 9,97 is the point on the axis of the abscissae (derived from the MATLAB® plot with the "data cursor" function) from which we start to see again the single dots of discrete values.

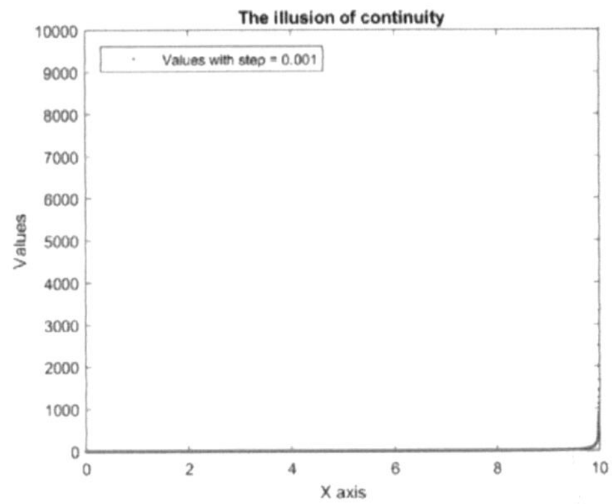

Fig. 5. The FM illusion of continuity for step Δx = 0.001 (see text).

The theoretical infinite value point is reached at x = 10, but also it slows the computation and it does not allow to reach the illusion of continuity for the last values. To solve this limitation, we should consider for the "Point D request" to vary the number of computed values in between every subset of the interval. The expected behavior is the need of higher steps in the firsts sub-intervals, while decreasing steeply towards the last ones. In this way the computation can be greatly reduced with respect to the continuous approach, and we can reach a better (or more far) illusion of continuity. The solution to this problem is presented in the following subsections.

4.2.3 Point C Result

A linear function is characterized a linear relation on the axis of the abscissae and of the ordinates. Thus, the

https://doi.org/10.1051/itmconf/20181601007

simplest solution for this request is plotting the values against the very same vector instead of the initial [0; 10] interval. This results in a linear behavior, that put into evidence the sparse points towards the right end of the representation window. The limitation on the illusion of the continuity, as you can see in Fig. 6, is much more evident with the linear behavior.

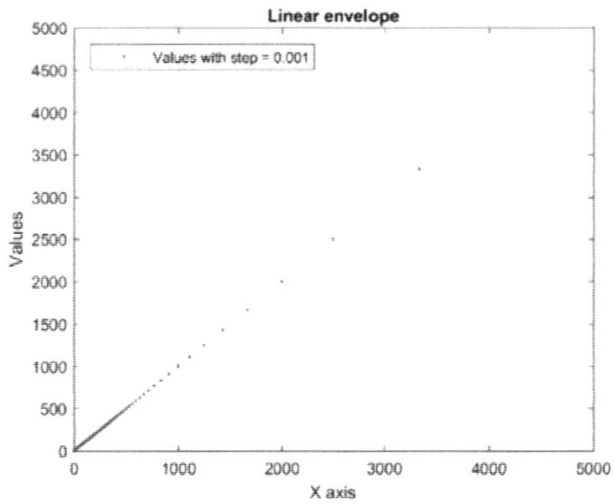

Fig. 6. Linear function representation of the FM with step Δx = 0.001 (see text).

This happens because the abscissae values become sparser with respect to the previous interval. In this configuration, unless limiting the interval to a smaller value (in this example, near to x = 400), it will be impossible, without further computation, to reach the illusion of a continuous plot. Anyway, with any linear behavior, a simple interpolation would be enough for the reconstruction of every missing point in the sparse regions.

4.2.4 Point D Result

As introduced in previous Point B subsection, in order to reduce the computational burden at each interval, while reaching the illusion of a continuous function, one approach can be adjusting the steps in each sub-interval inside the range [0; 10]. For this purpose, we create a different function called *Ex1b*. It is possible even to obtain a slightly more versatile function, allowing the user to define in which range he wants to create the exponential function. Therefore *Ex1b* accepts as input the value of this "range", defining the initial interval in [0; *range*]. Following the same approach of previous function *Ex1*, *Ex1b* defines three vectors. First, it generates an array that is not linear for the axis of the abscissae: based on previous remarks, it decreases the distance between the points as it proceeds to the right end of the interval. It is possible to define a general rule that is experimentally verified to give the illusion of a continuous curve along the entire representation interval. We can decrease the step between the points (from the starting 10^{-3} to the small value of 10^{-10}) every 4/5 of the remaining span, starting from the entire range and reaching the few points remaining near to the "range" value. This distribution of points compensates for the rarefaction effect due to the exponential shape of our initial FM. Then, the initial system values are computed in the same way implemented in *Ex1*: the results are shown in Fig. 7.

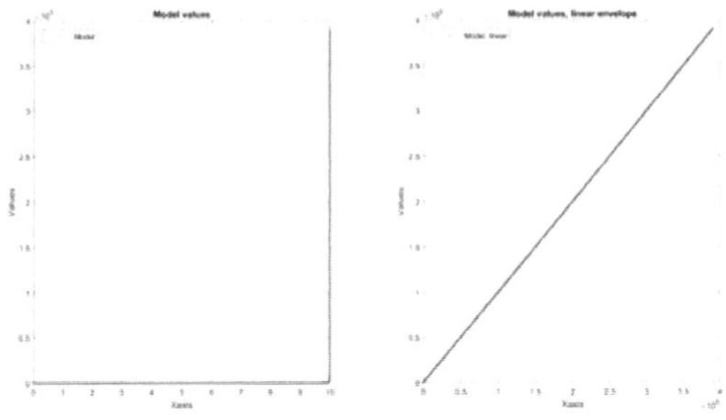

Fig. 7. FM with the illusion of continuity, both for the exponential and the linear behavior. Both graphics contain 204,000 points (see text).

We can notice that the function actually seems described by a continuous line, with respect to the resolution of the screen used (3200x1800 pixels), but with a far smaller computational effort with respect to the traditional approach. In fact, with a linear subdivision of the X axis as in Fig. 6, a plot with a range [0; 10] and a precision step of 10^{-10}, suitable for a continuous appearance at the end of the range, would require 10^{11} points. Instead, thanks to the discrete approach presented here, the final plot is constituted by just 204,000 points, with the same visual result. In other words, we obtain the reduction ratio RR = 99.9999796 % for representation information content savings.

When comparing Fig. 3, Fig. 5, and Fig. 7 we can notice that the envelope of the FM changes shape when increasing the number of points used, arriving to a theoretical orthogonal shape for precision $\rightarrow \infty$. This is due to the construction rules of the FM itself, which start dividing sequentially the values with respect to the X axis points. The final result, with an envelope similar as much as possible to the continuous line, visually, starts growing towards infinite very close to the right end of the representation window, and the envelope becomes closer to the ideal 90° angle (reaching the ideal orthogonality). In fact, at x = 9.99 we can see that the plot reaches a value of one thousand, while until x = 9.9 it remains under a hundred. In the first example instead, with just 10 values, we see a clear growing trend already at x = 2.0. Fig. 5 has a behavior that is in the half, with 10,000 points and a visible growing behavior after x = 9.5.

Furthermore, if we focus on the final envelope (Fig. 7 on the left) between the abscissae values 9.0 and 10.0, and we superimpose it to the initial 10.0 values shown in Fig. 2, we can notice that the behavior is the same. In Fig. 8 we show this comparison. On the left you can see the initial values showed in the interval [0; 9] (because the last value, corresponding to 10.0 is ∞), while on the right subplot there is the thickened line of the model, zoomed in between abscissae [9.92; 10.00] that reveals the same envelope with the illusion of continuity.

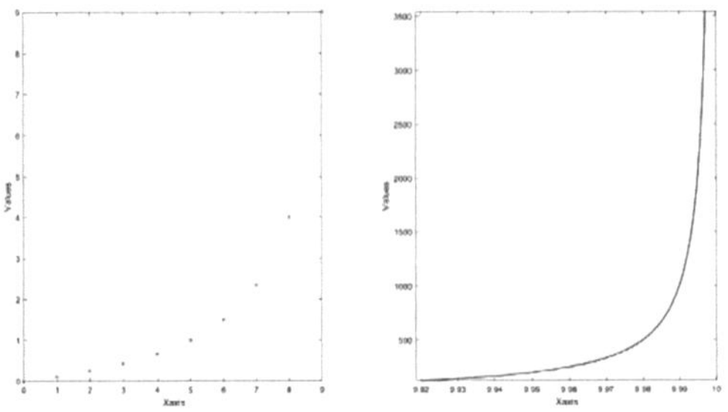

Fig. 8. Comparison of the initial values (on the left) and the corner of the thickened line (on the right).

We can conclude that, once reached the illusion of continuity, we obtained the maximum coverage on the used representation support for the function in the assigned representation window. Nevertheless, in order to achieve the minimum information content of the FM, we should assure that the number of abscissae points that represent the function are the minimum possible. In this example, instead, even if the total number of points was significantly reduced, obtained by function *Ex1b*, we cannot declare that this is the minimum information content of the FM that can be achieved, even if we obtained the reduction ratio RR = 99.9999796 % for representation information content savings!

5 Conclusion

In previous papers published elsewhere [7, 8, 10, 11, 15, 16, 21, 23, 24, 25, 26, 27, 28, 29, 30] we have already shown that traditional ***Q*** Arithmetic can be regarded as a highly sophisticated open logic, powerful and flexible bidirectional formal language of languages, according to "Computational Information Conservation Theory" (*CICT*) new perspective. We need to become deeply aware that human beings and most current advanced computational and measurement systems formal representations have quite a limited capacity to capture and to extract reliable information from noisy sources or generators traditionally. This is mostly due to the representation ambiguity at the core of traditional and current computational mathematics. This new awareness can offer competitive approach to guide more effective and convenient algorithm development and application to arbitrary multiscale (AMS) system modeling and simulation in advanced biomedical, sociology and social cybernetics applications. For *CICT* the well-posed arithmetic structural description is self-defined by its functional closures. This new awareness opens the way to new computational competence in computational science. In this paper we presented an articulated modeling OpeRational (OR) example to show how function model visual continuity appearance illusion can be achieved with big computational saving. Specifically we show how 99.9999796 % representation information content saving is obtained on a 3200x1800 pixels representation support (video screen). Science does not exists to enlighten people's minds only. It mainly exists to show the educated way from quanta to qualia. And that way starts from computational competence first.

References

[1] N. N. Taleb, R. Douady, *A Mathematical Formulation of Fragility* (Peer-Reviewed Monographs, Des Cartes, 2015).
[2] L. von Bertalanffy, *Perspectives on General System Theory* (Edited by Edgar Taschdjian, George Braziller, New York, 1974).
[3] F. Bailly, G. Longo, G., *Mathematics and the Natural Sciences. The Physical Singularity of Life* (Imperial College Press, London, 2011).
[4] L. A. Zadeh, Information and Control **8**, (1965).
[5] L. A. Zadeh, Information Processing 74, 3 (1974).
[6] I. Licata, in G. Minati, M. Abram, E. Pessa, Eds., *Methods, Models, Simulations and Approaches towards a General Theory of Change* (World Scientific, Singapore, 2012).
[7] R.A. Fiorini, in N.E. Mastorakis, M. Demiralp, N. Mukhopadhyay, F. Mainardi, Eds., *Advances in Applied mathematics, Modelling and Simulation, Mathematical and Computers in Science and Engineering Series n.34* (Florence, Italy, WSEAS Press, 385-394, 2014).
[8] R. A. Fiorini, Intntnl. J. Biol. Biom. Eng. **10** (2016).
[9] R. A. Fiorini, in G. Minati, M. Abram, E. Pessa, Eds., *Towards a Post-Bertalanffy Systemics* (Springer, New York, 2016).
[10] R. A. Fiorini, Intntnl. J. Syst. Applications, Eng. and Dev. **9**, 93 (2015).
[11] R. A., Fiorini, J. Soft. Science and Comp. Intel. **8**, 2 (2016).
[12] F. Geyer, Kybernetes **31**, 7/8 (2002).
[13] Y. Wang, B. Widrow, L.A. Zadeh, N. Howard, S. Wood, V.C. Bhavsar, G. Budin. C. Chan, R.A. Fiorini, M.L. Gavrilova, D.F. Shell, Intntl. J. Cog. Inf. and Nat. Intel **10**, 4 (2016).
[14] M. Bunge, *Emergence and Convergence* (Toronto, University of Toronto Press, 2003).
[15] R. A. Fiorini, J. Soft. Science and Comp. Intel. **9**, 1 (2017).
[16] R. A. Fiorini, Fund. Inf. **141** (2015).
[17] E. Kaltofen, in B. Buchberger, R. Loos, G. Collins, Computer Algebra (Springer Verlag, 1982).
[18] GMP (2017). Available at:<URL= https://gmplib.org/>.
[19] T. C. Tao, van H. Vu, *Cambridge Studies in Advanced Mathematics 105* (Cambridge University Press, Cambridge, 2006).
[20] E. W. Weisstein, *Full Reptend Prime* (1999-2012). Available at:<URL= http://mathworld.wolfram.com/FullReptendPrime.html>.
[21] R. A. Fiorini, G. Laguteta, Fund. Inf. **125** (2013).

[22] D. M. Young, R. T. Gregory, A Survey of Numerical Mathematics (vol.I and II, Reading, MA., USA, Addison Wesley 1973).
[23] R. A. Fiorini, G. F. Santacroce, in *Proceedings of International Symposium on Valencia 2013 -- The Economic Crisis: Time for a Paradigm Shift -- A Need for Paradigm Shift* (València, Spain, B.S.LAB, January 24-25, 2013).
[24] R. A. Fiorini, G. Dacquino, in K. L. Priddy (Ed.), SPIE Proc., *Intelligent Computing: Theory and Applications III* (SPIE vol. 5803, 2005).
[25] R. A. Fiorini, J. Soft. Science and Comp. Intel. (to be published).
[26] R. A. Fiorini, Fund. Inf. **135** (2014).
[27] R. A. Fiorini, in *Proc. Fourth International Conference on Mathematics and Computers in Sciences and in Industry, MCSI 2017* (Corfu Island, Greece, WSEAS Press, August 24-26. 2017).
[28] R. A. Fiorini, in J. Horne, Ed., *Philosophical Perceptions on Logic and Order* (AKATM Book Series, Hershey PA, USA, IGI Global, 2017).
[29] R. A. Fiorini, in N.E. Mastorakis, Gen Qi Xu, Eds., *Proceedings of the 3rd International Conference on Applied and Computational Mathematics (ICACM '14)* (NAUN Conferences, WSEAS Press, Geneva, Switzerland, December 29-31, 2014).
[30] R. A. Fiorini, in Proc. International Conference Applied Mathematics, Computational Science and Systems Engineering (AMCSE 2017) (Athens, Greece, October 6-8, 2017). To be published.

On the analysis of the limited resources queuing system under MAP arrivals

Eduard Sopin[1,2*] and *Konstantin* Samouylov[1]

[1]Peoples' Friendship University of Russia (RUDN University), 6 Miklukho-Maklaya St, Moscow, 117198, Russian Federation
[2]Institute of Informatics Problems, FRC CSC RAS, 44-2 Vavilova Str., Moscow, 119333, Russian Federation

Abstract. In the paper, we analyse a multiserver queuing system with discrete limited resources and random resource requirements under MAP arrivals, which can adequately model resource allocation schemes in the contemporary wireless networks. The equilibrium system of equations is derived in the vector form and is solved numerically. With stationary probability distribution, we provide formulas for the average and the variance of the occupied resources, as well as for the blocking probability. The results are illustrated by a numerical example.

1 Introduction

Fast mobile connections over the last decade become more available for business and personal communications. Since then portable devices have dramatically improved the processing capabilities and the battery saving. Both of this factors impact on the mobile data traffic and shift the focus from voice services to the video and data services.

The significant increase in the user traffic demands requires a broadband improvement and advances techniques in the future evolution of mobile systems, often referred to as fifth generation (5G) networks [1-2]. Among the implemented improving in modulation and coding, MIMO techniques, the network solutions are also targeted to improve the wireless network performance. Network densification [3], direct in-band and out-of-band device-to-device communications [4] are considered the popular and cost effective solutions to challenge the spatial frequency reuse and substantial capacity gain.

Application of a classic queuing models fails to cover a specific mechanism of the frequency resources allocation in LTE-Advanced networks. The physical resources is usually represented by a time-frequency resource grid. Each resource element is made of one OFDM subcarrier during the interval of one OFDM symbol. These resource elements can be grouped into resource blocks (RBs) with a total bandwidth of 180 kHz per one block. Based on the imbedded channel quality feedback mechanism each block of coded bits is matched with a target rate then a modulation scheme is selected. In a simple words RBs are scheduled in respect to the channel quality indicator (CQI) and required bitrate that a user equipment communicate with a base station before the beginning of each subframe (1 ms) [5]. The maximum achievable bitrate for a user session i can be provided by a Shannon formula:

$$C_i^{\max} = \omega \log_2(1 + \mathrm{SNR}_i\, p_{\max}),$$

where ω is a spectral bandwidth, $p_{\max}$ is a base station maximum transmit power and SNR_i is a signal-to-noise ratio for an i-th user session. It basically depends on the conditions of the radio signal propagation, the distance to the base station and its transmit power. The new session admission is allowed if $\sum_i \frac{C_i}{C_i^{\max}} \le 1$, where C_i is a required rate at each subframe that depends on a distance. Thus, the resource requirements of each user session depends on a number of model parameters and varies for each session. The ratio $\frac{C_i}{C_i^{\max}}$ may be interpreted as a share of system resources required by a user session.

The described above resource allocation process can be modelled in terms of multiserver queuing systems, in which customers require not only a server, but also a random amount of limited resources. Such queuing systems with random resource requirements, Poisson incoming flow and exponentially distributed service time was investigated in [6-8] and was extended in [9] by implementing signals that trigger resource reallocation process. However the real traffic model could hardly be adequately described by a Poisson arriving process. In this paper we consider a queuing system with random resource requirements under Markovian arrival process that can approximate a wide range of real processes. The proposed method of analysis is applied for only discrete resource requirements, the case of continuous resource requirements is left for further research.

* Corresponding author: sopin_es@rudn.university

The rest of the paper is organized as follows. Section 2 introduces the mathematical model in terms of queuing system with resource requirements of a random volume and mathematical framework for the performance measures evaluation. Numerical results are presented in section 3 while section 4 concludes the paper.

2 Mathematical model

In this section, we describe the queuing system with random resource requirements under MAP arrivals and provide the analysis of its stationary characteristics.

2.1 Model description

Consider a multiserver queuing system with N servers and R resources. Customers arrive according to a Markovian Arrival Process (MAP) governed by a continuous-time finite-state Markov chain with M states, which is named here the underlying Markov chain. The arrival process is characterized by the matrices $\mathbf{A}_0$ and $\mathbf{A}_1$, $\mathbf{A}_0+\mathbf{A}_1=\mathbf{A}$, where $\mathbf{A}$ is the generator matrix of the underlying Markov chain [10-11]. Assume that the matrix $\mathbf{A}$ is irreducible and denote θ the stationary probability vector of the underlying Markov chain. Then, the customer arrival rate is given by $\lambda=\theta\mathbf{a}$, where $\mathbf{a}=\mathbf{A1}$ and $\mathbf{1}$ is a vector of ones with appropriate size.

An arriving customer requires not only a server, but also j resources with probability p_j, $j\ge 0$. We assume the resource requirements to be mutually independent identically distributed random variables, independent of arrival and service processes. If upon arrival of a customer there are no available servers of not enough free resources in the system to meet the requirements, then the customer is lost. If upon arrival of a customer there are enough resources, then the customer occupies one server and the required number of resources for whole service duration. Service times are assumed to be exponentially distributed with intensity μ.

In order to simplify the analysis, we do not keep track of number of resources occupied by each customer. Instead of it, we keep track of only total number of occupied resources by all customers together [6]. Thus, we do not know the exact number of resources released on the departure of a customer. If there are k customers in the system just before the departure and r resources are occupied, then j resources are released upon the departure of a customer with probability $\dfrac{p_{r-j}^{(k-1)}p_j}{p_r^{(k)}}$, where $p_r^{(k)}$ is k-fold convolution of probability distribution $\{p_j\}$, $j\ge 0$.

The system behavior is described by a Markov process $X(t)=\{\xi(t),\delta(t),\eta(t)\}$, where $\xi(t)$ denotes number of customers in the system at time t, $\delta(t)$ – total number of occupied resources and $\xi(t)$ is the state of underlying Markov chain. The state set $X=\bigcup_{k=0}^{N}X_k$ can be easily decomposed into subset $X_0=\{(k,m):0\le k\le N,1\le m\le M\}$, and N subsets $X_k=\{(k,r,m):0\le k\le N,0\le r\le R,p_r^{(k)}>0,1\le m\le M\}$, $1\le k\le N$. Let s_k be the number of nonzero probabilities in the k-fold convolutions $p_r^{(k)}$, $0\le r\le R$. Then the number of states in subset X_0 is M, while subset X_k contain s_kM states. Finally, let $I(k,r)$ be the sequence number of $p_r^{(k)}$ at the row of nonzero k-fold convolutions of probability distribution $\{p_j\}$, $j\ge 0$.

2.2 Stationary probability distribution and performance measures

Denote stationary probabilities

$$q_0(m)=\lim_{t\to\infty}P\{\xi(t)=0,\eta(t)=m\},1\le m\le M\ , \quad (1)$$

$$q_k(r,m)=\lim_{t\to\infty}P\{\xi(t)=k,\delta(t)=r,\eta(t)=m\}, \\ (k,r,m)\in X_k. \quad (2)$$

Then vectors $\mathbf{q}_k=\left(\mathbf{q}_k(r_1),\ldots,\mathbf{q}_k(r_{s_k})\right)$, $1\le k\le N$ where $\mathbf{q}_k(r_i)=\left(q_k(r_i,1),\ldots,q_k(r_i,M)\right)$, $(k,r_i,m)\in X_k$, and $\mathbf{q}_0=\left(q_0(1),q_0(2),\ldots,q_0(M)\right)$ satisfy the following system of equilibrium equations.

$$\mathbf{q}_0\mathbf{D}_0+\mathbf{q}_1\mathbf{M}_1=\mathbf{0},$$
$$\mathbf{q}_{k-1}\mathbf{\Lambda}_{k-1}+\mathbf{q}_k\mathbf{D}_k+\mathbf{q}_{k+1}\mathbf{M}_{k+1}=\mathbf{0}\ ,1\le k\le N-1, \quad (3)$$
$$\mathbf{q}_{N-1}\mathbf{\Lambda}_{N-1}+\mathbf{q}_N\mathbf{D}_N=\mathbf{0}\ .$$

The matrix $\mathbf{D}_0$ is given by

$$\mathbf{D}_0=\mathbf{A}-\left(\sum_{j=0}^{R}p_j\right)\mathbf{A}_1, \quad (4)$$

while matrices $\mathbf{D}_k, 1\le k\le N$ have block-diagonal structure. The dimension of the matrices $\mathbf{D}_k, 1\le k\le N$ is $s_kM\times s_kM$ and the size of each block is $M\times M$. Diagonal blocks $\mathbf{d}_k(j,j), 1\le j\le s_k$ of $\mathbf{D}_k, 1\le k\le N-1$ are described by formula (5) :

$$\mathbf{d}_k\left(I(k,r),I(k,r)\right)=\mathbf{A}-k\mu\mathbf{I}-\left(\sum_{j=0}^{R-r}p_j\right)\mathbf{A}_1, \\ 0\le r\le R,\ p_r^{(k)}>0, \quad (5)$$

where $\mathbf{I}$ is a unit matrix. All diagonal blocks of $\mathbf{D}_N$ are calculated by

$$\mathbf{d}_N\left(I(N,r),I(N,r)\right)=\mathbf{A}-N\mu\mathbf{I}. \quad (6)$$

The matrix $\mathbf{\Lambda}_0$ can be represented as a row of $M\times M$ size blocks, the blocks $\mathbf{\lambda}_0(i), 1\le i\le s_1$ have the following form

$$\mathbf{\lambda}_0\left(I(1,r)\right)=p_r\mathbf{A}_1,\ 0\le r\le R,\ p_r>0. \quad (7)$$

The size of matrices $\mathbf{\Lambda}_k$, $1 \le k \le N-1$ is $s_k M \times s_{k+1} M$, their blocks $\boldsymbol{\lambda}_k(i,j)$, $1 \le i \le s_k$, $1 \le j \le s_{k+1}$ are given by

$$\boldsymbol{\lambda}_k\left(I(k,r), I(k+1,s)\right) = p_{s-r}\mathbf{A}_1, \quad 0 \le r \le s \le R,\ p_r^{(k)} > 0,\ p_s^{(k+1)} > 0. \tag{8}$$

The matrix $\mathbf{M}_1$ can be represented as a column of blocks of size $M \times M$, the blocks $\mathbf{m}_1(i), 1 \le i \le s_1$ have the following form

$$\mathbf{m}_1\left(I(1,r)\right) = \mu\mathbf{I},\ 0 \le r \le R,\ p_r > 0. \tag{9}$$

The size of matrices $\mathbf{M}_k$, $2 \le k \le N$ is $s_k M \times s_{k-1} M$, their blocks $\mathbf{m}_k(i,j)$, $1 \le i \le s_k$, $1 \le j \le s_{k-1}$ are given by

$$\mathbf{m}_k\left(I(k,r), I(k-1,s)\right) = \frac{p_{r-s} p_s^{(k-1)}}{p_r^{(k)}} k\mu\mathbf{I}, \quad 0 \le s \le r \le R,\ p_r^{(k)} > 0,\ p_s^{(k-1)} > 0. \tag{10}$$

The system of equations (3) along with normalizing condition gives stationary probabilities (1-2).

With the stationary probability distribution, we can evaluate various performance metrics of the system, such as the average b_1 and the variance $\overline{b}_2$ of occupied resources :

$$b_1 = \sum_{(k,r,m)\in X_k} r q_k(r,m), \tag{11}$$

$$\overline{b}_2 = \sum_{(k,r,m)\in X_k} r^2 q_k(r,m) - b. \tag{12}$$

According to the properties of MAP, the blocking probability π is given by

$$\pi = 1 - \lambda^{-1}\left(\sum_{\substack{k,r:0\le k\le N-1 \\ 0\le r\le R, p_r^{(k)}>0}} \mathbf{q}_k(r) \sum_{j=0}^{R-r} p_j \right)\mathbf{a}. \tag{13}$$

3 Numerical analysis

In this section, we provide some results of the performance measures analysis. Mapping radio resource allocation policy to the probability distribution $\{p_j\}$, $j \ge 0$ of the resource requirements in the queuing system is a quite complicated problem [12], so we left it to the further research.

We assume that resource requirements follow geometric distribution with parameter $p = 0.8$ and

$$p_j = (1-p)p^j, \quad j \ge 0.$$

Calculations were held for three arrival flows with equal arrival intensity ($\lambda = 8$) but different variation coefficient v of the interarrival times. The first arrival flow with $v = 0.7$ is defined by

$$A = \begin{bmatrix} -6 & 3 & 3 \\ 3 & -5 & 2 \\ 3 & 5 & -8 \end{bmatrix}, \quad A_1 = \begin{bmatrix} 1 & 3 & 3 \\ 3 & 0 & 8 \\ 2 & 2 & 0 \end{bmatrix}.$$

Second arrival process represents MMPP flow (Markov Modulated Poisson Process) with variation coefficient $v = 1.5$:

$$A = \begin{bmatrix} -4 & 1 & 3 \\ 12 & -24 & 12 \\ 3 & 1 & -4 \end{bmatrix}, \quad A_1 = \begin{bmatrix} 1 & 0 & 0 \\ 0 & 8 & 0 \\ 0 & 0 & 15 \end{bmatrix}.$$

And third arrival process is a Poisson flow with variation coefficient of the interarrival times $v = 1$.

Further, we assume that $N = R = 40$ and provide evaluation of performance metrics (11 – 13) for different values of system load $\rho = \dfrac{\lambda}{\mu}$. The results are plotted on figures 1, 2 and 3.

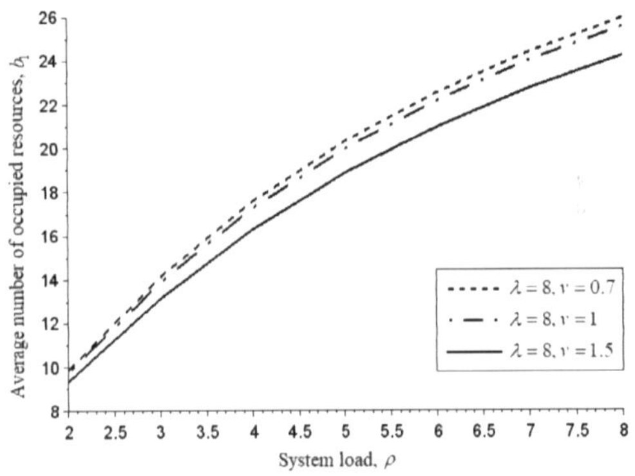

Fig. 1. Average number of occupied resources.

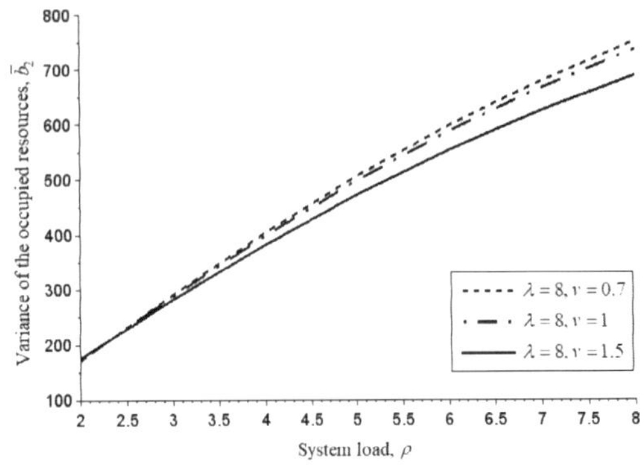

Fig. 2. Variance of the occupied resources.

Figures show that all of the investigated performance indicators improve with the decrease of the variability of interarrival times. Since the majority of request flows in real wireless networks have coefficient of variation less than one, assumption of arrival flow to be distributed according to Poisson law leads to lower bound of performance measures. However, to receive more accurate results one need to approximate the arrival flow with MAP.

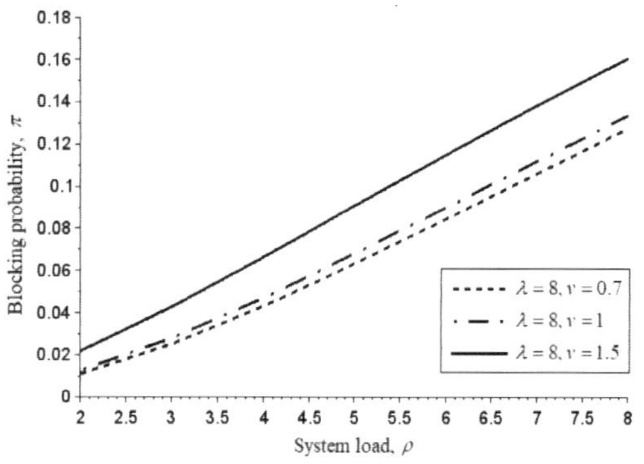

Fig. 3. Blocking probability.

4 Conclusion

In the paper, we analysed a multiserver queuing system with discrete limited resources under MAP arrivals. The described queuing system may be applied for the analysis of performance measures of modern wireless networks. We derived the system of equilibrium equations that can be solved numerically. Formulas for the main performance measures were deduced. Finally, we illustrated our results by numerical example.

In the future studies, we plan to develop an efficient algorithm for approximation of performance measures.

The research was financially supported by the Ministry of Education and Science of the Russian Federation (the Agreement number 02.a03.21.0008), and RFBR according to the research projects No. 16-37-60103 and No. 16-07-00766.

References

1. Ericsson mobility report, Ericsson AB, 2017 https://www.ericsson.com/assets/local/mobility-report/documents/2017/ericsson-mobility-report-june-2017.pdf
2. Cisco Visual Networking Index: Global Mobile Data Traffic Forecast Update, 2016-2021 https://www.cisco.com/c/en/us/solutions/collateral/service-provider/visual-networking-index-vni/complete-white-paper-c11-481360.pdf
3. J. Andrews, H. Claussen, M. Dohler, and S. Rangan, *IEEE Journal on Selected Areas in Communications*, **30**, 3, pp. 497–508 (2012).
4. G. Fodor, S. Parkvall, S. Sorrentino, P. Wallentin, Q. Lu, and N. Brahmi, *IEEE Access*, **2**, 1, pp. 1510–1520 (2014).
5. J. Fan, Q. Yin, G. Li, B. Peng, and X. Zhu, *IEEE Transactions on Wireless Communications*, **10**, 11, pp. 3966–3972 (2011).
6. O.M. Tikhonenko, *Automation and Remote Control*, **58**:6, pp. 969–973 (1997).
7. V.A. Naumov, K.E. Samuilov, A.K. Samuilov, *Automation and Remote Control*, **77**, 8, pp 1419–1427 (2016).
8. V. Naumov, K. Samouylov, N. Yarkina, E. Sopin, S. Andreev and A. Samuylov, *Proc. of 7th Congress on Ultra Modern Telecommunications and Control Systems*, pp. 100–103 (2015).
9. K. Samouylov, E. Sopin and O. Vikhrova, Analysis of the queueing system with resources and signals, *Proc. of 8th Congress on Ultra Modern Telecommunications and Control Systems*, (in print)
10. G. Basharin and V. Naumov, *Proc. of the 3rd International Seminar on Teletraffic Theory "Fundamentals of Teletraffic Theory"*, Moscow, VINITI, pp. 38–44 (1984).
11. D. M. Lucantoni, K. S. Meier-Hellstern, and M. F. Neuts, *Advances in Applied Probability*, **22**, 3, pp. 676–705 (1990).
12. Sopin E., Samouylov K., Vikhrova O., Kovalchukov R., Moltchanov D., Samuylov A., *Lecture Notes in Computer Science*, **9870**, pp. 440–450 (2016).

https://doi.org/10.1051/itmconf/20181602001

Analysis of Relationship Between Personality and Favorite Places with Poisson Regression Analysis

Ha Yoon Song, Hwa Baek Kang

Department of Computer Engineering,
Hongik University, Seoul, Korea

Abstract A relationship between human personality and preferred locations have been a long conjecture for human mobility research. In this paper, we analyzed the relationship between personality and visiting place with Poisson Regression. Poisson Regression can analyze correlation between countable dependent variable and independent variable. For this analysis, 33 volunteers provided their personality data and 49 location categories data are used. Raw location data is preprocessed to be normalized into rates of visit and outlier data is prunned. For the regression analysis, independent variables are personality data and dependent variables are preprocessed location data. Several meaningful results are found. For example, persons with high tendency of frequent visiting to university laboratory has personality with high conscientiousness and low openness. As well, other meaningful location categories are presented in this paper.

1. Introduction

Location Based Service (LBS) is one of the emerging topic with wide possibility of future service. Especially, understanding human mobility pattern is one of the core part of LBS.

In addition, it is widely recognized that human personality may affect personal favorite location. The relationship between human personality and human preferred location is valuable to understand.

In this paper, we are going to analyze the relationship between human personality and human favorite location by regression analysis. The personality data will be independent variable and the location data will be treated as dependent variable.

Linear regression is a general tool to analyze the correlation between independent variable and dependent variable. However, normal linear regression model such as stepwise regression cannot be a general tool in our case.

In our paper, the independent variable is from Big Five Factor (BFF) of human personality and the dependent variable is from the count of visit to favorable location categories. Since linear regression model is not an adequate to analyze such count data, we will use Poisson regression to analyze data set containing count data.

Poisson regression model is based on Poisson distribution. Poisson distribution is a probability distribution regarding the count of occurrences in a given unit of time or area which probability mass function is shown as equation (1).

$$p(y) = \lim_{n\to\infty} \begin{bmatrix} n \\ y \end{bmatrix} p^{y}(1-p)^{n-y} = \frac{\lambda^{y}}{y!} e^{-\lambda} \quad (1)$$

Where y is a variable standing the number of occurrences, n is total number of events, and λ is expectation of probability variable y. Poisson distribution is a convergence of binomial distribution with n goes to infinity.

The structure of this paper is like the followings. Section 2 discuss about the related researches. In section 3, we will discuss about the details of data used in this paper. In section 4, the pretreatment of data for Poisson regression will be discussed. Section 5 will present the result of Poisson regression and analyze the result. In section 6, we will conclude this paper and discuss about possible future research topics.

2. Previous Studies

Recently, positioning data can be collected by many of commercial devices including smartphones. Thus, the mobility data can be easily collected and these data can be used for various research purpose. For example, user can check in their favorite places voluntarily.

The question arise that human personality and human favorite locations are related. Since these data

Corresponding author: hayoon@hongik.ac.kr

are personal data and privacy protection required, rare previous works and be found.

I. Burbey, [1] tried to predict next location based on past movement pattern. S. Y. Kim and H. Y. Song [2] used Back Propagation Network (BPN) to analyze mobility data. P. T.

Costa and R. R. McCrae [3] categorize personality into five element: openness, conscientiousness, extraversion, agreeableness, neuroticism. It is possible to enumerate each personality elements so that the corresponding numbers can be directly utilized in our research.

The most related work to this study can be found in H. Y. Song and E. Y. Lee [4], which used Stepwise regression to analyze the relationship between personality and visiting place. Stepwise regression is repetitive iteration of regression analysis, and is know to find the relationship between dependent variable and independent variable. However, stepwise regression cannot analyze relationship between countable data while the frequency to visit a location is clearly a count data. Therefore, we are going to use Poisson regression to analyze the relationship in this paper.

3. Personality Data and Location Data

3.1 Personality Data

For personality data in our research, FFM(Five Factor Model) is utilized. Big Five Factor (BFF) of a person can be obtained by survey using Big Five Inventory (BFI) and 33 volunteers provided their BFF. BFF is composed of five personality factor openness, conscientiousness, extraversion, agreeableness and neuroticism L. R. Goldberg, [5].

The advantage of BFF is that the personality factors are presented in numerical form between 0 to 5 and these values can be independent variables of regression analysis.

Table 1. Personality Data

Person \ BFF	O	C	E	A	N
person1	3.3	3.9	3.3	3.7	2.6
person2	2.7	3.2	3.3	2.7	2.8
person3	4.3	3.1	2.3	3.2	2.9
person4	3.6	3.3	2.8	3.2	2.8
person5	4.2	4.3	3.5	3.6	2.6
person6	4.0	3.7	4.0	3.9	2.8
person7	3.5	3.8	3.4	3.2	3.0
person8	2.2	3.4	3.0	3.1	2.6
person9	2.6	2.8	3.4	3.1	2.6
person10	3.3	2.9	3.1	3.1	3.3
person11	3.4	3.2	3.4	3.3	3.1
person12	3.1	3.7	3.4	3.2	3.5
person13	3.4	3.6	3.6	2.9	2.5

·O: Openness

·C: Conscientiousness

·E: Extraversion

·A: Agreeableness

·N: Neuroticism

Table 1 shows part of BFF of total 33 volunteers BFI survey. BFFs of thirteen volunteers are presented. Therefore, direct comparison of personality can be enabled in quantitative manner. For example, person1 has higher Openness than person2 and has lower Neuroticism than person2.

3.2 Location Data

Location data can be collected by checking-in user favorite places. For such purpose, smartphone app called SWARM [6] is used by volunteers. Table 2 shows parts of location data collected. It contains place names, place categories, and the count of visit to the categories for person2.

The topmost count shown in Table 2 is 69 at university laboratory by person2. The university laboratory is categorized as university building. The count to a category is a sum of count to places belonging to the category.

Table 2. Location Data

Count	Category	Place name
69	University building	University laboratory
36	Korean restaurant	Seorae Babsang
16	Pastfood restaurant	Burger king
11	Book store	Hongik book store
4	Theater	CGV hongik

For example, student restaurant, university library, and so on belong to category of university building and the counts of student restaurant, university library, university laboratory is summarized into the count of university building category.

Once the category count of a specific place is less than or equal to five, the place is categorized into etc. category. It is because to avoid the effect of minute count to regression analysis. In sun, we have 49 categories used in this paper.

4. Data Preprocessing

Table 4. Raw Location Data

	airport	church	Train station	Outdoor Activity	library	University laboratory	Bus station	Shopping mall	Subway station	home
person1	23	5	15	72	422	107	6	11	7	375
person2	17	3	1	23	306	23	0	201	1	341
person3	7	4	2	2	84	11	0	5	2	79
person4	11	0	2	20	515	13	0	1	0	516
person5	5	2	0	9	270	10	1	4	0	169
person6	5	0	1	3	257	33	4	6	4	121
person7	5	38	0	4	199	46	11	2	39	145

Table 5. Pruned and Normalized Location Data

	airport	church	Train station	Outdoor Activity	Library	University laboratory	Bus station	Shopping mall	Subway station	home
person1	0.014715	0.003199	0.009597	72	0.269994	0.068458	0.003839	0.007038	0.004479	0.239923
person2	0.013127	0.002317	0.000772	23	0.236293	0.017761	0	0.155212	0.000772	0.26332
person3	0.014737	0.008421	0.004211	2	0.176842	0.023158	0	0.010526	0.004211	0.166316
person4	0.005624	0	0.001022	20	0.263292	0.006646	0	0.000511	0	0.263804
person5	0.007163	0.002865	0	9	0.386819	0.014327	0.001433	0.005731	0	0.24212
person6	0.006297	0	0.001259	3	0.323678	0.041562	0.005038	0.007557	0.005038	0.152393
person7	0.007418	0.05638	0	4	0.295252	0.068249	0.01632	0.002967	0.057864	0.215134
person8	0.005814	0	0.004651	10	0.196512	0.001163	0.002326	0.016279	0	0.276744

The counts of location categories are utilized as dependent variable of regression analysis. However, several preprocessing stages required in advance to the regression analysis. The counts need to be pruned. The raw count may effect to regression analysis negatively with biases. Therefore, too high count and too low count must be excluded before regression analysis. Out of 36 data set of volunteers, 2 set of data are excluded because of their too high count and one set of data is excluded because of its too low count.

In addition, the counts need to be normalized. Therefore, we normalized the raw counts in t rates by equation (2).

$$\boldsymbol{Rate} = \frac{visiting\ count\ by\ category}{total\ visiting\ count} \quad \textbf{(2)}$$

The normalized data can be used as dependent variable.

5. Regression Analysis Result

Poisson regression is the major tool we used for the analysis of relation between location data and personality data.

Calibrated count of location categories in Table 4 is used as dependent variable, and personality data in Table 1 is used as independent variable. The Poisson regression library incorporated in R studio is used. Figure 1 shows a sample analysis results executed in R studio environment. Poisson regression analysis is done for the location category 'theater' and personality data.

The main result in Fig 1 shows the effect of five independent variables, O, C, E, A, and N toward the location category theater. First, results with p-value less than 0.05 must be observed. This is because that, statistically, result of regression analysis with p-value judged to be meaningful. In this example, the visit to theater and factor O (Openness) is related. The values in Estimate column is quantitative effect of independent variable to dependent variable.

 https://doi.org/10.1051/itmconf/20181602001

Fig 1. Poisson Regression Example

The estimate of O in Figure 1 is 2.46986, which leads to analysis result that a person with high Openness is tent to visit theater frequently. On the contrary, once we have negative values on Estimate for the effect of one of BFF to a location category, it will be judged that low BFF value leads to high tendency to visit a location.

Table 6 shows summarized results of total regression analysis. There are 49 location categories to be analyzed and 23 meaningful location categories are found. For each location categories, Poisson regression analysis is done and found meaningful when p value of each test found less than 0.05. And estimate value of each BFF is recorded in Table 6.
Results column in Table 6 shows the symbols of the effective personality factor. The sign '+' indicates positive effect. Once we have estimate values greater effect of C to the location category university laboratory.

On the contrary, O (Openness) has negative effect

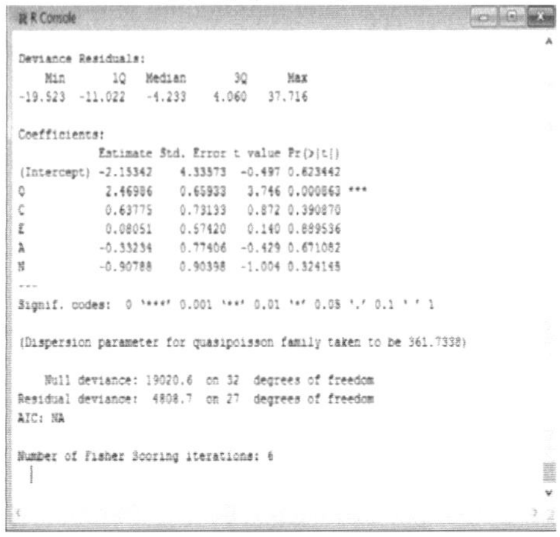

Table 6. Regression An alysis Result

BFF / Location	Openness O	Conscientiousness C	Extraversion E	Agreeableness A	Neuroticism N	Result
University Building	-1.34122		0.33211		2.30918	-O, ++N
University Laboratory	-2.111	2.80482	1.20598		-0.22139	--O, ++C, +E
Library	1.33241	1.222134		-2.11323		+O, +C, --A
Theater	2.46986					++O
Korea Restaurant		0.221422	-1.22311			-E
Airport	2.13322			-1.22332	0.33224	++O, -A
Church		1.22342		3.222114	0.592213	+C, +++A,
Train station	-1.22322	1.98273		0.22111		-O, +C
Bank		0.29223		-0.23324	0.333221	
Subway station	1.29878	-0.11111				+O
Resort			-0.22211		0.44221	
Government Office	0.12498			0.332211	-1.22334	-N
Shopping mall		1.928722		1.222552		+C, +A
Bus station	0.33429		-0.45438		0.391248	
Hospital	0.11234			-2.31429	1.09932	-A, +N
Rest area		-0.33245			1.32222	+N
Gym	1.43222					+O
Book store				0.32222		
Outlet	0.11582		0.33222			
Home	-1.82522			1.21121		-O, +A
Museum	1.53222					+O
China Restaurant		1.43222			-0.32221	+C
Italy Restaurant	-1.32222			0.44223		-O

than 2, the sign '++' is used. On the contrary, the sign '-' indicates negative effect between personality and location category, and the sign '--'-rsonality and locateffect when the estimate values is less than -2. A blank cell in Table 6 stands for meaningless results from Poisson regression analysis.

There is interesting tendency shown in Table 6. Poisson regression analysis indicates that University Laboratory are positively related with C (Conscientiousness) having Estimate value of 2.80482 and thus indicated as ++C for the location category university laboratory, meaning high positive with estimate value of -2.111 on University Laboratory and indicated as --O. As a result, a person with high conscientiousness and low openness has strong tendency to visit university laboratory frequently.

6. Conclusion

In this study, we tried to analyze a relationship between frequently visiting place of a person and the person's personality by use of Poisson regression. Personality can be represented by BFF, and frequent locations can be obtained by Swarm app and preprocessed.

Since the frequentness of visit is represented by count of visit, Poisson regression is regarded as an adequate method. In addition, we categorized and normalized location data.

Comparing to results of previous research by H. Y. Song and E. B. Lee [4], our result shows similarity in several locations. For example, University Building shows -O,++N in our research and very similar to that of previous research which shows --O for University Building. The difference maybe due to various reasons such as data used, regression method, and so on while the trend of relationship is still valid. As a result of this study, several location categories are found to have meaningful relations with BFF.

Also our research can be applied to many of related areas in order to increase imposed value of each area, such as enhancement of mobility model D. Alberg, M. Last, and S. Elnekave, [7], and D. Guo and W. Cui, [8].

We hope our research results can be applied to various location based service. For example, more prediction rate can be found in case user's personality is incorporated in location prediction system.

There are lots of analysis method left untried, such as ZINB regression, spearman rank correlation coefficient, and so on, and applying other methods on our dataset can reveal latent relationships or find the relationship more clearly.

Acknowledgement

This work was supported by the National Research Foundation of Korea (NRF) grant funded by the Korea government (MEST) (NRF-2017R1D1A1B03029788) and (NRF-2016K1A3A7A08952097)

References

1. I. Burbey, "Predicting future locations and arrival times of individuals,"2011
2. S. Y. Kim and H. Y. Song, "A study on influence of human personality to location selection, 2016, pp. 267-285.
3. P. T. Costa and R. R. McCrae, "Four ways five factors are basic, d Rrsonality and individual differences, **vol. 13, no. 6,** 1992, pp. 653–665.
4. H. Y. Song and E. B. Lee, An Analysis of the Relationship between Human Personality and Favored Location
5. L. R. Goldberg, "The structure of phenotypic pe rsonality traits." American psychologist**, vol. 48, no. 1,** 1993, p. 26.
6. Swarm Application, https://www.swarmapp.com
7. D. Alberg, M. Last, and S. Elnekave, "A spatio-temporal simulation model for movement data generation," in Data Mining Workshops, 2008. ICDMW'08. IEEE International Conference on. IEEE, 2008, pp. 320–325.
8. D. Guo and W. Cui, "Mining moving object trajectories in location based services for spatio-temporal database update," in Sixth International Conference on Advanced Optical Materials and Devices. International Society for Optics and Photonics, 2008, pp. 71 432M–71 432M.

Influence of slope and external temperature on smoke stratification in case of fire in bi-directional road tunnel

Peter Weisenpacher[1,*], *Jan* Glasa[1], and *Lukas* Valasek[1]

[1]Institute of Informatics, Slovak Academy of Sciences, Dubravska cesta 9, 84507 Bratislava, Slovakia

Abstract. Jet fan ventilation strategy in case of fire in bi-directional road tunnels is focused on maintaining smoke stratification. There are several factors influencing stratification under specific conditions. In this paper smoke movement during a 5 MW fire in a 600 m long road tunnel is studied by computer simulation and the influence of slope and external temperature on smoke stratification is analysed. Calculations were performed on a high performance computer cluster using parallel version of Fire Dynamics Simulator. Smoke stratification upstream of the fire is maintained in every simulation scenario with the exception of declivous tunnel, in which buoyancy intensifies backlayering. The behaviour of the smoke movement downstream of the fire is more complex. In the case of horizontal tunnel the stratification is not maintained in the vicinity of the fire and region with untenable conditions expands downstream. In the tunnel with slope of -2° this expansion is accelerated, while in the tunnel with slope of 2° untenable conditions spread in opposite direction. The influence of exterior temperature higher than temperature inside the tunnel is relatively weak in horizontal tunnels; however, it becomes very important in sloping tunnels, especially downstream of the fire.

1 Introduction

In case of fire in bi-directional road tunnels people are trapped both upstream and downstream of the fire. Therefore, tenability conditions for human life should be maintained long enough to ensure safe evacuation in both parts of the tunnel tube. An air flow with critical velocity pushing smoke away from the tunnel cannot be used as the conditions downstream of the fire would become untenable for people. Instead, jet fans are used to achieve the prescribed target velocity of the flow maintaining stratification of smoke layer. Although backlayering cannot be prevented under such circumstances, the smoke layer spreads above human heads and tenable conditions for human life are preserved.

An important question connected with this strategy is the determination of a proper value of target velocity. Various national regulations recommend the values of velocity ranging from 1.0 to 1.5 $m.s^{-1}$ (for example [1, 2]). After the prescribed target velocity is achieved, the flow interacts with buoyancy forces created by the fire as well as with specific tunnel tube geometry. It may influence air flows in some regions of the tunnel as well as spread of smoke. Therefore, a thorough analysis is needed to evaluate specific fire scenarios. The main mechanism causing the descent of smoke layer is entrainment between the hot and cold layer [3]. However, there is a lack of knowledge for the general applicability of entrainment equations [4, 5]. Therefore, the results of computer simulation of specific fire and ventilation scenarios can become a useful guide for tunnel management. It can be also used for training and education of tunnel specialists [6, 7]. This problem is of special importance in Slovakia, where several road tunnels with longitudinal ventilation are currently under construction, including some bi-directional tunnels [8, 9].

In [10] we have investigated smoke movement in a 600 m long road tunnel by computer simulation for fire scenarios with various heat release rates. The influence of several prescribed values of the target air flow velocity on smoke stratification and visibility has been examined for chosen placement of the fire source. The simulation results indicate that it is difficult to maintain smoke stratification for fires with heat release rate (HRR) exceeding the value of 3 MW corresponding to passenger car fire. There has been no value of the target velocity maintaining smoke stratification in both upstream and downstream parts of the tunnel found, although the 1 $m.s^{-1}$ velocity leads to the highest visibility achieved in evaluated sectors of the tunnel. Velocity above 1 $m.s^{-1}$ produces smoke eddies in the downstream part of the tunnel, while lower values of velocity result in significant backlayering upstream of the fire.

In this study, the smoke stratification in a 600 m long road tunnel is evaluated for selected fire scenarios with the 5 MW HRR and the air flow velocities of 1.0 and 1.5 $m.s^{-1}$ simulated by the Fire Dynamics Simulator (FDS), version 6.5.2 [11,12]. An influence of external temperature and the tunnel slope on maintaining smoke

* Corresponding author: upsyweis@savba.sk

stratification is studied. Smoke movement as well as tenability criteria for the considered scenarios are evaluated in particular parts of the tunnel tube. As our research has demonstrated the applicability of parallel versions of the FDS for tunnel fire simulations [13], calculations are performed on the high-performance computer (HPC) cluster.

2 Fire Dynamics Simulator

The FDS is a CFD-based simulation system for modelling fire and fire-driven fluid flows. I was developed by the National Institute of Standards and Technology (USA) with cooperation with VTT Technical Research Centre of Finland (Finland) [11,12]. The FDS numerically solves a form of conservation equations for low-speed, thermally-driven flows with an emphasis on the smoke and heat transport from fire. It includes models of fire related physical and chemical processes such as turbulence, thermal radiation, pyrolysis, combustion of the pyrolysis products, conductive heat transfer, etc. The FDS enables various kinds of parallelization of calculation.

Input data required for the FDS simulation include geometry of the investigated compartment (the tunnel size and shape, location, size and shape of all objects in the tunnel), representation of jet fans, material properties (ignition temperature, thermal conductivity, specific heat, density, emissivity, heat of evaporation, etc.) and the fire HRR. The most important output quantities of the gaseous component are temperature, velocity, pressure and density of the gas, the concentration of smoke and visibility.

3 Fire Scenarios

We use the FDS model of the 600 m long bi-directional road tunnel. The tunnel has circular cross section and is equipped with one pair of jet fans (see Fig. 1).

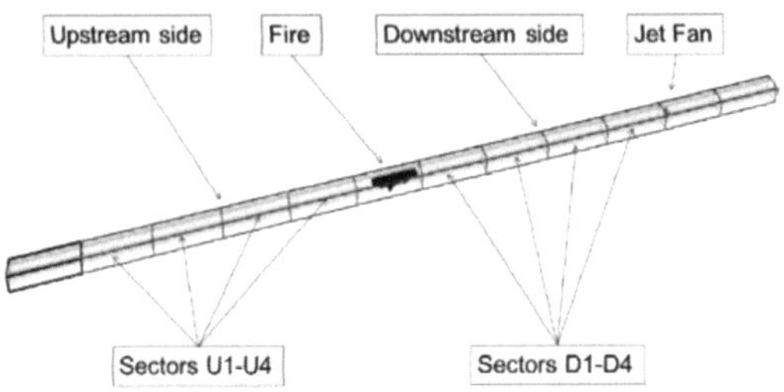

Figure 1. Scheme of the tunnel.

Dimensions of the cross section of the tunnel are 10.8 m width, 6.8 m height. Cross section area of jet fans is 0.5 m^2, their length is 3.7 m. These parameters are typical for tunnels which are currently under construction in Slovakia. The distance of jet fans pair from the left portal is 503.2 m. The tunnel includes two regions in which stratification is evaluated (on both sides of the fire). Each of them is approximately 200 m long. The length of tunnel region necessary for dissipation of jet fan airflow may exceed 100 m. Therefore, the length of the tunnel is sufficient to describe all considered phenomena related with smoke stratification realistically.

The fans are represented as an HVAC, using rectangular vents of dimensions 0.6 m x 0.8 m with prescribed normal velocity. The volume flow of the jet fans at the beginning of the simulation is 15 $m^3.s^{-1}$. After the required target velocity of the air flow in the tunnel is reached, the performance of jet fans decreases with the aim to stabilize the flow velocity at a prescribed value. Two different values of the target velocity are tested: 1.0 and 1.5 $m.s^{-1}$. The initial value of air velocity in the tunnel is 0 $m.s^{-1}$.

The fire source is modelled by a rectangular block placed 276 m from the left tunnel portal. Its top surface (4 m^2) burns with the heat release rate per unit area (HRRPUA) of 1250 $kW.m^{-2}$ producing the 5 MW fire. The fire HRR increases linearly, reaching its maximal value at the 40th s of the fire. The fire soot yield is 0.2.

The smoke distribution is evaluated in eight sectors of the 50 m length (see Fig. 1). Four of them (U1 – U4) are located in the upstream part of the tunnel, other four in the downstream part (D1 – D4). Somewhere in this area emergency exits would be situated in a real tunnel; therefore, maintaining the tenability conditions for human life [14, 15] in these sectors would be the aim of tunnel management. Tenability conditions require the visibility at human head level above 10 m. In each sector the visibility at head level (1.6 m) averaged within its whole area is evaluated during the 420 s fire. This quantity provides only simplified aggregated measure of smoke distribution; however, it describes main trends of the smoke movement sufficiently.

Several fire scenarios are investigated for two values of target velocity. In the scenarios *V1.0* and *V1.5* for the target velocity of 1.0 and 1.5 $m.s^{-1}$, respectively, the empty horizontal tunnel and the 10°C ambient temperature within the tunnel are assumed. In the scenario *V1.x-C* 20 passenger cars are distributed on both sides of the fire. In the scenario *V1.x-E*, the influence of modification of external temperature to 15°C is studied. The scenario *V1.0-Es* with the same parameters includes also the simulation of 2000 s before the start of the fire in order to achieve the realistic stabilized velocity profile induced by different internal and external temperatures at the beginning of the fire. In the *V1.x-Sl-2* and *V1.x-Sl2* scenarios, the slope of the tunnel is -2° and 2°, respectively. In the latter one the opposite direction of jet fan flow is necessary to vanquish buoyancy forces created by the fire. Therefore, the different location of the jet fans (48 m from the left portal) is used to prevent a direct interaction of their jets with smoke. The *V1.x-Sl2E* and *V1.x-Sl2Es* scenarios evaluate the influence of the 15°C external temperature for the tunnel with 2° slope. In the latter scenario, a stabilization of the air flow in the tunnel before the fire start is also simulated.

The computational domain size is 600 m x 12 m x 7.2 m. For the 20 cm mesh resolution, the domain consists of 3000 x 60 x 36 cells. The total number of

Scenario	U1 [m]	U2 [m]	U3 [m]	U4 [m]	D1 [m]	D2 [m]	D3 [m]	D4 [m]	Jet [m/s]	Vel [m/s]
V1.0	30,0	30,0	30,0	29,3	4,7	7,9	15,0	21,0	11,0	1,0
V1.0-C	30,0	30,0	30,0	30,0	4,8	6,0	14,1	22,8	13,1	1,0
V1.0-E	30,0	30,0	28,0	26,7	3,8	6,4	16,3	23,9	10,9	1,0
V1.0-Es	30,0	30,0	28,0	26,9	4,1	7,0	15,1	20,7	11,1	1,0
V1.0-Sl2	30,0	30,0	30,0	30,0	23,2	28,6	5,5	1,3	-8,0	1,0
V1.0-Sl2E	30,0	30,0	30,0	29,7	24,4	18,8	2,0	1,1	-11,1	1,0
V1.0-Sl2Es	30,0	30,0	30,0	29,8	26,0	14,5	2,4	1,1	-12,6	1,1
V1.0-Sl-2	29,8	24,5	7,4	2,2	1,1	0,9	0,9	0,9	21,4	1,0
V1.5	30,0	30,0	30,0	30,0	3,2	3,1	3,6	3,8	14,1	1,5
V1.5-C	30,0	30,0	30,0	30,0	4,7	4,6	3,8	3,9	16,7	1,5
V1.5-E	30,0	30,0	30,0	29,8	3,0	2,9	3,3	3,7	14,1	1,5
V1.5-Es	30,0	30,0	30,0	29,8	2,6	2,9	3,2	3,5	14,2	1,5
V1.5-Sl2	30,0	30,0	30,0	30,0	7,5	9,1	12,0	20,7	6,2	1,5
V1.5-Sl2E	30,0	30,0	30,0	30,0	6,3	8,2	10,7	19,1	-3,8	1,5
V1.5-Sl2Es	30,0	30,0	30,0	30,0	6,1	8,4	14,9	18,1	-6,9	1,5

Figure 2. Average visibility in the evaluated sectors U1 – U4 and D1 – D4 in the upstream and downstream part of the tunnel during the last 20 s of the fire; jet fan air flow velocity and velocity averaged within the whole tunnel cross section for all considered scenarios.

cells is 6,480,000. In order to increase the simulation performance the parallel MPI version 6.5.2 of the FDS is used. The computational domain was decomposed into 12 meshes, each of them assigned to one MPI process (one CPU core). Eight of these meshes correspond to sectors U1 - U4 and D1 - D4.

All simulations were carried out on the SIVVP HPC cluster at the Institute of Informatics, Slovak Academy of Sciences, Bratislava (Slovakia) [16]. It is an IBM dx360 M3 cluster consisting of 54 computational nodes (23 Intel E5645 @ 2.4 GHz CPU, 48 GB RAM); the total number of cores is 648. The nodes are connected by the Infiniband interconnection network with the bandwidth of 40 Gbit/s per link and direction. The total CPU time depends on specifications of each fire scenario. The CPU time varies from 79 hours for the scenario *V1.0-Sl2* to 278 hours (*V1.5-C*).

4 Simulation Results

The results of the simulations are consistent with [10]. Fig. 2 shows average visibility in the evaluated sectors in the upstream and downstream region of the tunnel during last 20 s of the fire for all considered scenarios. The colours of the cells indicate tenable or untenable conditions during considered period. It can be seen that the effect of passenger cars placed in the tunnel as well as the effect of different exterior temperature on the visibility in horizontal tunnel is relatively weak. However, the effect of the slope as well as contribution of the temperature differences between internal and external temperature to this effect are considerable.

4.1 Smoke stratification upstream of the fire

A significant backlayering occurs in all tested scenarios, although the length of backlayering strongly depends on simulation parameters. However, the smoke stratification upstream of the fire is well maintained in every simulation scenario with the exception of the scenario with the -2° tunnel slope. Smoke layer upstream of the fire for selected scenarios can be seen in Fig. 3.

As can be expected, higher target velocities result in shorter lengths of backlayering. Moreover, in the scenario *V1.0* a slight drop of visibility occurs, while in *V1.5* a faster air flow maintains a perfect visibility. The cars located in the tunnel have a similar effect. As the average velocity within the tunnel cross section at the beginning of the tunnel is kept constant, the flow velocity above cars roofs must be increased to fulfil the mass conservation equation, which decelerates the movement of the smoke layer in opposite direction. Faster air flow also prevents a slight decrease of visibility which can be observed in scenario V1.0. The effect of temperature differences between internal and external temperatures is more pronounced than Fig. 2 indicates. Warmer air flowing into the tunnel decreases the temperature gradient between hot and cold layer dissolving smoke layer and increasing entrainment. However, a small drop of visibility in these scenarios does not constitute a threat for people, as tenability conditions are fulfilled.

In the simulations *V1.x-Sl2* buoyancy forces suppress backlayering even more while in the scenario *V1.x-Sl-2* backlayering is increased. In the latter scenario buoyancy increases the velocity of smoke layer. As the average velocity within the tunnel cross section is constant, the velocity of flow above the road in opposite direction must be increased as well, thus intensifying entrainment, dissolving smoke layer and disrupting smoke stratification. This is the reason of a lower visibility in this scenario. In two of the four evaluated sectors (U3 and U4) the visibility drop causes conditions untenable for human lives.

Figure 3. Smoke layer upstream of the fire after 400 s for the scenarios *V1.0, V1.0-C, V1.0-E, V1.0-Es, V1.0-Sl2, V1.0-Sl2E, V1.0-Sl2Es, V1.0-Sl-2* and *V1.5*.

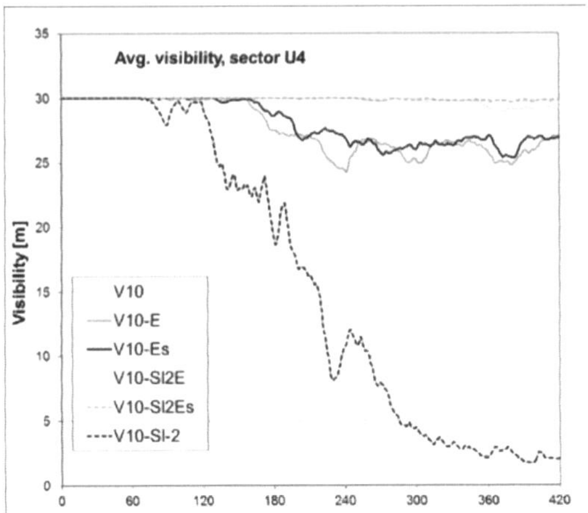

Figure 4. Time behaviour of averaged visibility in the sector U4 for selected scenarios

Differences between scenarios with or without the 2000 second stabilisation period are very small. The influence of natural flow of warmer air inside the tunnel at the beginning of the fire in the scenarios with stabilization is negligible in comparison with the influence of warm air flowing into the tunnel as a consequence of jet fan activity.

In all tested scenarios in which the visibility drop is observed it occurs in sectors U3 and U4. Time behaviour of averaged visibility in critical sector U4 for scenarios where perfect visibility upstream of the fire is not maintained can be seen in Fig. 4. Initial drop of visibility occurs after 2-3 minutes to the fire. Such behaviour is consistent with the results of [10].

In the same way, the higher target velocity is more efficient in maintaining the smoke stratification upstream of the fire. Only very small visibility decrease occurs in the case of some scenarios with the 1.5 $m.s^{-1}$ target velocity in sector U4.

4.2 Smoke stratification downstream of the fire

None of the tested values of the target velocity in any scenario is able to maintain smoke stratification in all evaluated sectors downstream of the fire (see Fig. 2). However, time behaviour as well as the final values of the averaged visibility in the sectors D1 - D4 are very different for particular scenarios (see Figures 5 – 8).

The slope of the tunnel has the greatest impact on smoke movement. The drop of visibility and descent of the smoke layer in horizontal tunnel typically starts in the vicinity of the fire (sector D1) due to interaction of the fire with the air flow and spreads downstream to the other sectors [10]. The target velocities higher than 1.0 $m.s^{-1}$ intensifies this effect (see Figures 2 and 5). This pattern can be seen also in Fig. 6. In the sectors D3 and D4 tenability conditions are fulfilled until the end of the simulation, unlike the scenario V1.5.

Figure 5. Smoke layer downstream of the fire after 150 s for scenarios *V1.0, V1.0-C, V1.0-E, V1.0-Es, V1.0-Sl2, V1.0-Sl2E, V1.0-Sl2Es, V1.0-Sl-2* and *V1.5*.

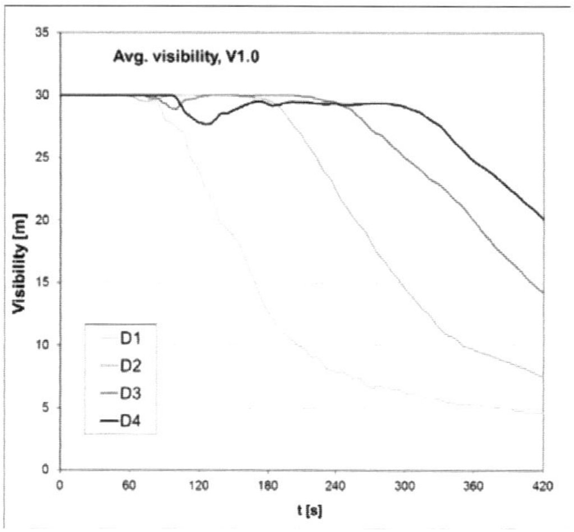

Figure 6. Time behaviour of averaged visibility in sectors D1 - D4 in scenario *V1.0*.

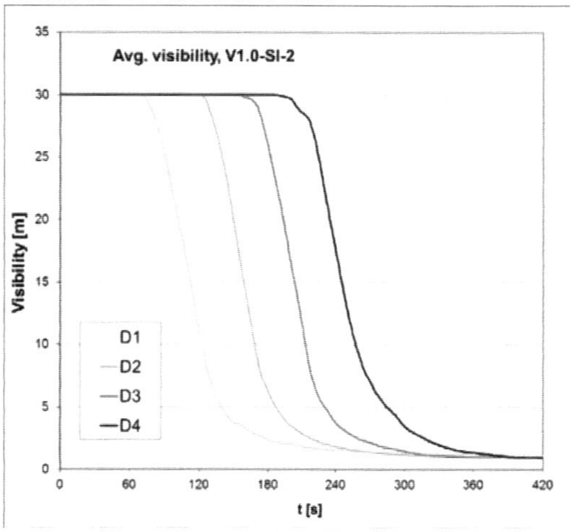

Figure 7. Averaged visibility in scenario *V1.0-Sl-2*.

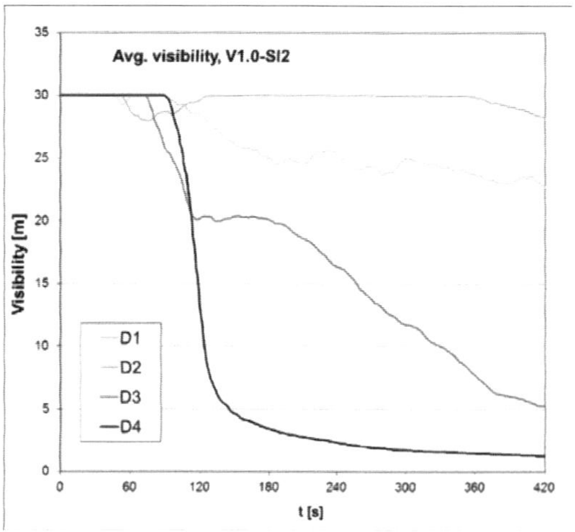

Figure 8. Averaged visibility in scenario *V1.0- Sl2*.

In the tunnel with the slope of -2° visibility drops faster than in the horizontal tunnel under the same circumstances. Until the 5th minute of the fire conditions for human life become untenable in all the sectors (see Fig. 7). The reason of this behaviour is buoyancy which increases an air flow velocity in front of the fire. It has a similar effect as a significant increase of target velocity. In such conditions the smoke stratifications cannot be maintained.

In the tunnel with the slope of 2° the smoke behaviour is more complex (see Fig. 8). There are two mechanisms decreasing visibility. In the sector D1 it is a direct interaction of the air flow with the fire. It is the same mechanism as in the scenarios discussed above; however, it is attenuated due to the effect of buoyancy which decreases the air velocity in front of the fire. This is the reason of tenability conditions in the sectors D1 and D2 being well-preserved until the end of the simulation. However, buoyancy increases the movement of smoke under the tunnel ceiling downstream of the fire and thus also entrainment. It results in a considerable dissolution of the smoke layer in the sector D4 and in somewhat less pronounced dissolution in the sector D3. Tenability conditions in these sectors are not fulfilled at the end of simulation. The slope of the tunnel changes the course of the smoke layer dissolution completely.

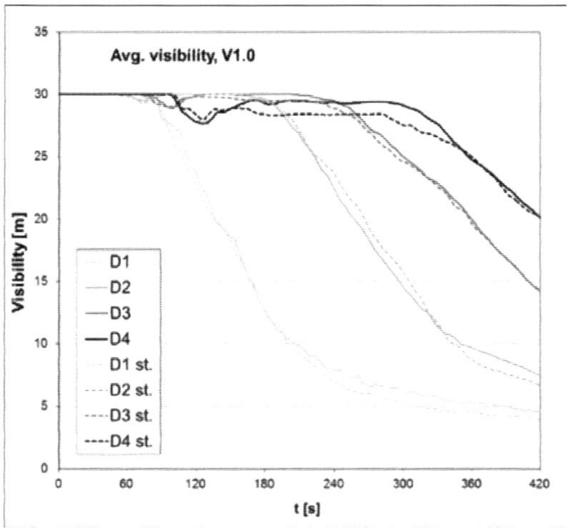

Figure 9. Averaged visibility in scenario *V1.0* (with and without stabilisation).

Although the effect of exterior temperature in horizontal tunnel is relatively weak (see Fig. 9), there is a considerable impact of the temperature gradient on smoke movement in sloping tunnel (see Fig. 10). In *V1.0-Sl2Es* scenario a combination of the slope and temperature gradient creates a flow of warmer air under the tunnel ceiling at the beginning of the fire, which does not exist in non-stabilized scenario. Although the higher temperature under the ceiling decelerates dissolution of the smoke layer, it simultaneously increases the difference between the velocities under the ceiling and above the road surface and thus it increases the entrainment.

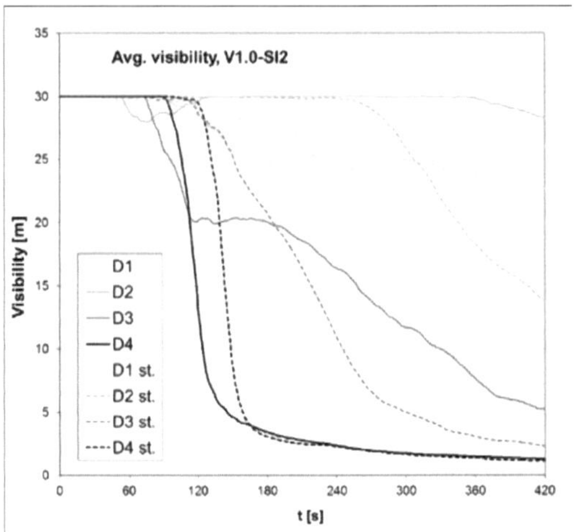

Figure 10. Averaged visibility in scenario *V1.0-SI2* (with and without stabilisation).

In the early phase of the burning, the first of above mentioned mechanisms prevails in the sectors D3 and D4 resulting in a better visibility in the stabilized scenario. After approximately 3 minutes of the fire the influence of natural warm flow becomes negligible in comparison with the heat produced by the fire and the effect of a more intensive entrainment prevails. The visibility in stabilized scenarios becomes worse. In the sector D2 the latter mechanism prevails during the whole simulation. The sector D1 is the only one in which the visibility in the stabilized scenario is better than in the non-stabilized.

5 Conclusions

In this paper the smoke movement during 5 MW tunnel fire is simulated by the FDS, version 6.5.2 for selected fire scenarios. The influence of slope and external temperature on the smoke stratification is investigated for two different values of the target air flow velocity.

The smoke stratification upstream of the fire is well maintained in every simulation scenario with the exception of the scenario with the -2° tunnel slope, in which buoyancy strongly increases backlayering. The influence of temperature differences on the smoke stratification is relatively weak. The first drops of visibility typically start in the sectors in the vicinity of the fire.

The most important factor influencing stratification downstream of the fire is slope. In horizontal tunnel the visibility drop starts in the vicinity of the fire and spreads downstream. In the tunnel with the slope of -2° this spread is accelerated, while in the tunnel with the slope of 2° the region with untenable conditions expands in the opposite direction. If the exterior temperature is higher than the temperature inside the tunnel, visibility is deteriorated even more significantly.

Acknowledgements

The authors would like to thank P. Schmidt, B. Spisakova (National Motorway Company) for information about technical specifications of road tunnels and M. Dobrucky, J. Astalos and V. Sipkova for technical support for SIVVP HPC cluster computing. This paper was partially supported by the Slovak Science Foundation (project No. VEGA 2/0165/17) and the Slovak Research and Development Agency (project No. APVV-15-0340).

References

[1] TP 12/2011 Road tunnels ventilation (in Slovak), (Ministry of Transport and Construction of the Slovak Republic, 2011)
[2] Guideline - Road tunnels ventilation (in Czech), (Ministry of Transport of the Czech Republic, 2013)
[3] H. Ingason, Y.Z. Li, A. Lönnermark, *Tunnel fire dynamics* (Springer-Verlag, New York, 2015, p. 321)
[4] R.L. Alpert, *Fire induced turbulent ceiling-jet* (Report no. 19722-2. Factory Mutual Research Corp., Norwood, 1971)
[5] H.Z. You, G.M. Faeth, Fire and Materials. **9**, 46 (1985)
[6] P. Danisovic, J. Sramek, E. Jancarikova, *Proceedings of 9th annual International Conference of Education, Research and Innovation,* Seville, Spain (2016)
[7] E. Jancarikova, P. Danisovic, J. Sramek, *Proceedings of 5th International Conference "Integration, Partnership and Innovation in Construction Science and Education"*, Moscow, Russia (2016)
[8] http://ndsas12.cmsromboid.sk/prehlad-tunelov/44346s
[9] http://www.sta-ita-aites.sk/index.php?page=tunely_vo_vystavbe
[10] P. Weisenpacher, J. Glasa J, L. Valasek, *Proceedings of 10th Mediterranean Combustion Symposium,* Naples, Italy (in press)
[11] K. McGrattan, S. Hostikka, R. McDermott, J. Floyd, C. Weinschenk, K. Overholt, *Fire Dynamics Simulator, Technical Reference Guide* (National Institute of Standards and Technology, Gaithersburg, Maryland, USA, and VTT Technical Research Centre of Finland, Espoo, Finland, sixth edition, 2017)
[12] K. McGrattan, S. Hostikka, R. McDermott, J. Floyd, C. Weinschenk, K. Overholt, *Fire Dynamics Simulator, User's Guide* (National Institute of Standards and Technology, Gaithersburg, Maryland, USA, and VTT Technical Research Centre of Finland, Espoo, Finland, sixth edition, 2017)
[13] P. Weisenpacher, J. Glasa, L. Halada, L. Valasek, V. Sipkova, Comput. Inform. **33** 1237 (2014)
[14] *Practice Note for Tenability Criteria in Building Fires* (Society of Fire Safety, Engineers Australia, 2014)
[15] D.A.Purser, *Assessment of Hazards to Occupants from Smoke, Toxic Gases and Heat* (In: P. J. DiNenno (Ed.), The SFPE Handbook of Fire Protection

Engineering. 4th Edition, National Fire Protection Association, Quincy, MA 02269, 2009, pp. 2-96 - 2-193)
[16] SIVVP – Slovak Infrastructure for High Performance Computing. Available from: http://www.sivvp.sk/, http://hpc.ui.savba.sk/

https://doi.org/10.1051/itmconf/20181602003

CFD Analysis of 2D Unsteady Flow Past a Square Cylinder at Low Reynolds Numbers

Zhenquan Li[1,*], and *Robert* Wood[1]

[1]School of Computing and Mathematics, Charles Sturt University, NSW 2640, Australia

Abstract. A study of the behaviour of flow past a square cylinder for Reynolds numbers 10 and 20 is presented. Open source software Navier2d in Matlab is used in this study. The investigation starts from a uniform initial mesh and then refine the initial mesh using a mesh refinement method which was proposed based on both qualitative theory of differential equations and the finite volume method implemented in Navier2d. The horizontal and vertical velocity component profiles and pressures are shown on the once refined meshes. The comparisons between the profiles and pressures are conducted to show the variations from Reynolds number 10 to 20. The twice refined meshes are also presented and these refined meshes provide the information where the behaviour of flow is complex.

1 Introduction

The study on wakes of flow around a bluff body has various applications in practice such as aerodynamics, wind engineering, and electronics cooling. The flow around a square cylinder has been investigated extensively using different methods in literature [e.g. 1, 2].

We use a different computational technique to analyse the flow past a square cylinder. A 2D adaptive mesh refinement method is proposed based on the qualitative theory of differential equations and utilise an existing numerical method [4]. The accuracy of the method is verified by analytical and CFD velocity fields considering the location of centres of vortices and/or separating curves of different vortex regions [5, 6, 8, 9]. The reliability of the method is considered and the positive outcome had been achieved [7]. The cost (computational complexity) is investigated using lid-driven cavity flow and a positive outcome again is achieved [10]. We use Navier2d as the numerical method. Navier2d is a vertex centred Finite Volume (FV) code that uses the median dual mesh for the Control Volumes (CV's) about each vertex. One of the advantages of the adaptive mesh refinement method is that a mesh is refined based on the information of the calculated velocity field on the mesh so the refined mesh shows where the velocity is more complex which is presented by more nodes in the refined mesh.

In this paper, we apply the adaptive mesh refinement once to the initial mesh. We show the velocity fields and pressures calculated on the once refined mesh, and the twice refined meshes that provide the information in the areas that contains centres of vortices and separation curves of vortex regions that are in the domain of interest [9].

2 The governing equations

In this paper, we use Navier-Stokes equations for unsteady flow of an incompressible fluid of density as governing equations

$$\frac{\partial \boldsymbol{u}}{\partial t} + \boldsymbol{u}\cdot\nabla\boldsymbol{u} = -\nabla p + \frac{1}{Re}\nabla^2\boldsymbol{u}$$
$$\nabla\cdot\boldsymbol{u} = 0 \qquad (1)$$

Where Re is the Reynolds number. The domain is shown in Fig. 1.

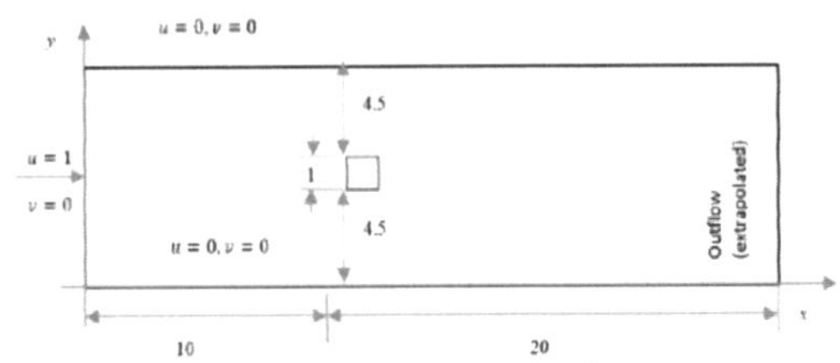

Fig. 1. Computational domain.

The domain in this study is a rectangle with the length 30 units and width 10 units. The unit square cylinder is placed inside of the rectangle with the centre located at (10.5, 5). The flow is from left to right. The origin of the xy coordinate system is located at the lower left-hand corner of the rectangle and the x axis is in the downstream direction.

The boundary condition at the inlet is u=1 and v=0. On the top and bottom length of the rectangle and the boundary of the square cylinder, u=v=0. The boundary

* Corresponding author: jali@csu.edu.au

condition at the outlet is the outflow extrapolate chosen in Navier2d.

3 Numerical results

We use uniform initial mesh with size 20×60. The initial mesh is refined based on the CFD velocity field calculated on the initial mesh. We then calculate the CFD velocity field on the once refined mesh. From the once refined mesh, we obtain the information about where the linearly interpolated velocity field does not follow the equivalent continuity equation [4]. The twice refined meshes are also shown. We consider the cases for Reynolds number Re=10 and 20. Fig. 2 shows the initial mesh.

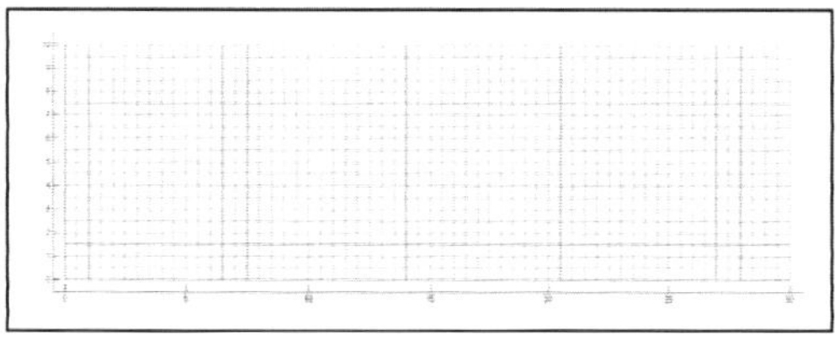

Fig. 2. Initial mesh used in study.

Navier2d requires the triangular mesh in the calculations. Fig.3 shows the triangular mesh generated from the initial mesh.

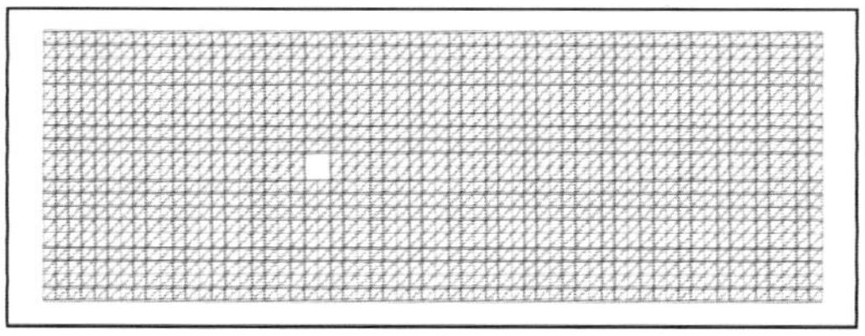

Fig. 3 Triangular mesh generated from the initial mesh.

Even though we start from the same initial mesh, the once and twice refined meshes are different due to different CFD velocity fields. The residuals in this study are less than 10^{-10}. The top figure shows the case Re=10.

3.1. Horizontal velocity comparison

The following figures present the differences between the u velocity component for the two cases Re=10 and 20.

From the two figures in Fig. 4, the wake of Re=10 is longer than that of Re=20 and the width of the wake of Re=10 is wider than that of Re=20. The flow gets more complex around the cylinder for Re=20 than that for Re=10 from both the once refined mesh and the profiles.

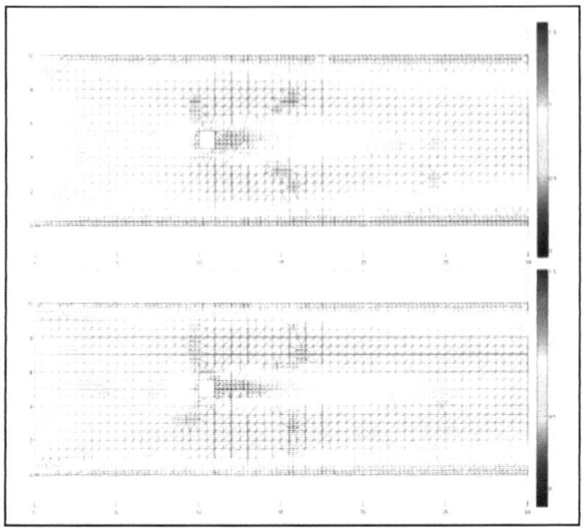

Fig. 4. The profiles of u components.

3.2 Vertical velocity comparison

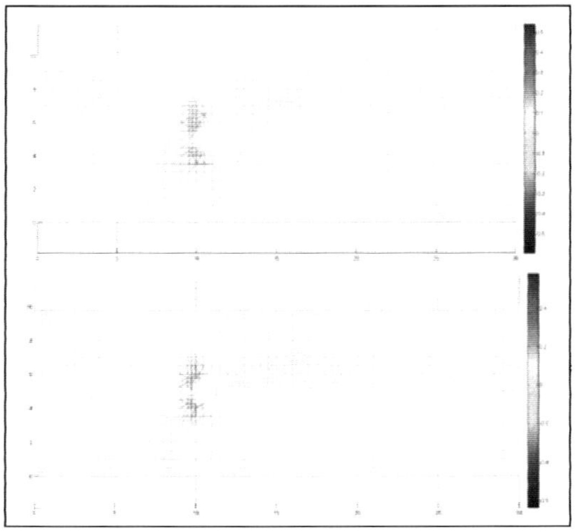

Fig. 5. The profiles of v components.

Fig. 5 shows the v profiles for both cases, in each case there are less differences than displayed in the u profile.

3.3 Pressure comparison

The comparison of pressures is shown in Fig. 6. There are large differences between the pressures. The pressure before the cylinder is larger for Re=10 than that of the case Re=20 since the flow is faster in the latter case.

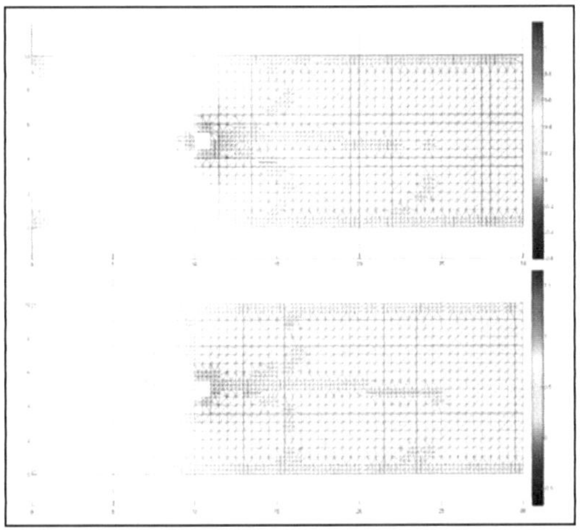

Fig. 6. Pressure profiles.

3.4 Twice refined meshes

The twice refined meshes are shown in Fig. 7 for both cases. In the region before the cylinder, the case Re=10 has a higher pressure so this area requires more fine grids as indicated in the top graph of Fig. 7. After the cylinder, the case Re=20 has a longer tail and the velocity field is sensitive in the vicinity of the tail. This is also indicated in the bottom graph of Fig. 7.

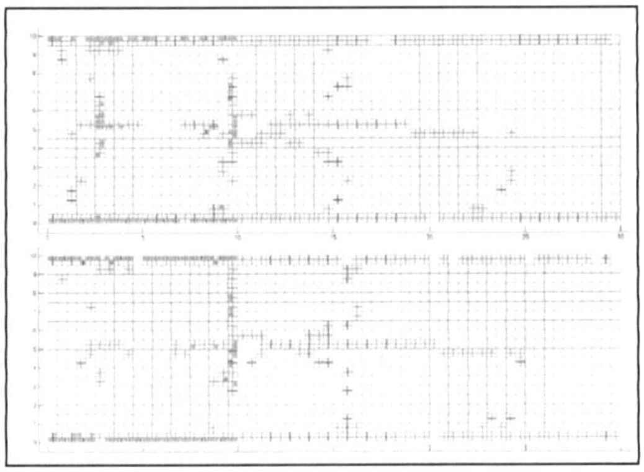

Fig. 7. Twice refined meshes.

3.5 Velocity fields

Fig. 8 shows the velocity fields calculated from Navier2d on the once refined mesh for both cases. For the case Re=10 in the top figure, the wake is shorter when compared with that of the bottom figure which is the case Re=20. There are larger differences between the two velocity fields after the cylinder. We also see the errors in these two figures. The domain is symmetrical, the boundary and initial conditions are the same for the two regions separated by y=5. However, the velocity fields are not symmetrical. Comparing the figures in Fig. 7 and Fig. 8, we conclude that the twice refined meshes are totally dependent of the numerical velocity fields calculated which is the same as we found before.

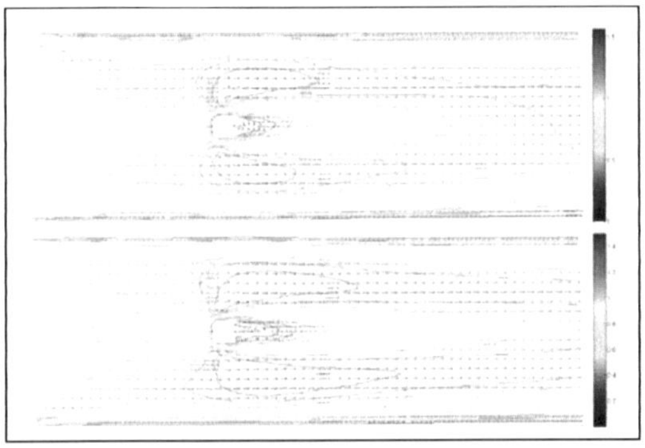

Fig. 8. Velocity fields.

4 Discussion

Even though the results from the uniform initial are not presented in this paper, our previous research has shown that the refined meshes generate more accurate results.

We have shown the simple cases in this paper. More complex cases are being investigated. The boundary conditions used in this paper are different from the boundary conditions used in [1, 2]. There are practical problems following the boundary conditions described in Section 2.

In the future, we will consider improving the accuracy of the numerical velocity fields in two ways. The first way is to choose finer initial mesh and the second is more refinements performed on the initial mesh.

References

1. S. Sen, S. Mittal, G. Biswas, Int. J. Numer. Meth. Fluids, **67,** 1160-1174(2011). DOI : 10.1002/fld.2416
2. B. Gera, P.K. Sharma, P.K. Singh, Int. J. Appl. Eng. Res., **1**, 602-610(2010). http://ipublishing.co.in/jarvol1no12010/EIJAER2028.pdf
3. D. Engwirda, *Navier–Stokes solver (Navier2d)*. MATLAB Central File Exchange, 2006.
4. Z. Li, J. Flow Visual. Image Process., **15**, (2008). DOI: 10.1615/JFlowVisImageProc.v15.i1.20
5. Z. Li, J. Flow Visual. Image Process., **13**, 1(2006). DOI: 10.1615/JFlowVisImageProc.v13.i1.10
6. Z. Li, J. Math. Chem., **52**, 4(2014). DOI : 10.1007/s10910-014-0334-0
7. R. Lal, Z. Li, J. Math. Chem., **53**, 3(2015). DOI: 10.1007/s10910-014-0461-7
8. Z. Li, J. Comput. Appl. Math., **275**, (2015). DOI: 10.1016/j.cam.2014.07.025
9. Z. Li, R. Wood, J. Comput. Appl. Math., **318**, (2017). DOI : 10.1016/j.cam.2016.09.022
10. Z. Li, Comput. Therm. Sci., to appear.

Integrated artificial intelligence algorithm for skin detection

Idoko John Bush[1,2,*], *Rahib Abiyev*[1,2], *Mohammad Khaleel Sallam Ma'aitah*[1] *and Hamit Altıparmak*[1]

[1]Computer Engineering Department, Near East University, Lefkosa, Northern Cyprus, Mersin 10, Turkey
[2]Applied Artificial Intelligence Research Centre, Near East University, Lefkosa, Northern Cyprus, Mersin 10, Turkey

Abstract. The detection of skin colour has been a useful and renowned technique due to its wide range of application in both analyses based on diagnostic and human computer interactions. Various problems could be solved by simply providing an appropriate method for pixel-like skin parts. Presented in this study is a colour segmentation algorithm that works directly in RGB colour space without converting the colour space. Genfis function as used in this study formed the Sugeno fuzzy network and utilizing Fuzzy C-Mean (FCM) clustering rule, clustered the data and for each cluster/class a rule is generated. Finally, corresponding output from data mapping of pseudo-polynomial is obtained from input dataset to the adaptive neuro fuzzy inference system (ANFIS).

1 Introduction

One of the unique features of humans is the skin. In several fields, skin detection has applications including identity confirmation and identification based on military and security applications, facial characteristics, camera remote control, video conferences, skin tracking in video, information access, managing pictorial information in banks and so on. This paper explores ANFIS algorithm to detecting skin from nonskin instances. Fuzzy neural networks and systems are generally known estimators that estimates any given non-linear function with intended accuracy under the condition of the presence of sufficient neurons in the middle layer and fuzzy rules. Recent studies in the field of fuzzy systems and neural networks indicate that the combination of these two methods can be very effective in nonlinear systems identification.

ANFIS structure works by integrating adaptive neural networks and fuzzy logic parameters which can be set based on the input and output data to model the systems [1]. The basic principle used in training the adaptive networks is the rule of reducing gradient. Using gradient reduction alone is not desirable due to the slowness and tendency of the local minima to set network parameters. However new hybrid methods that are the result of combining gradient reduction and other similar methods lead to increased speed of least square error (LSE). Fuzzy modeling and detection was earlier analyzed by Takagi-Sugeno and obtained practical applications in the field of control, detection and prediction. Adaptive neuro fuzzy inference systems works by combining fuzzy structures with artificial neural networks that are used to detect systems, predict time series and other various cases. There is a high level of uncertainty to achieve the desired automatic segmentation in each image segmentation layout. This fact also can be extended into face detection especially skin color segmentation. Therefore, fuzzy theory is a good way to achieve the basic detection because fuzzy model provides a mechanism that presents image uncertainty. According to the fuzzy rules extracted from different color spaces used in training phase, each pixel can be divided into the skin or non-skin pixels. One of the features of pixels-based color is that it required no space ground, therefore fixed and fast orientation and size for processing, detection and tracking applications such as detection of specified body parts of human, nudity detection and detection of face; all of which benefits from skin detection. Moreover, detection of skin color helps to block the abusive image or video content. In addition to its use in computing, color of human skin color plays a vital role in human and human relationships.

Dealing with black and white colors does not provide any satisfaction with image. Projecting the degree of membership is a new way to solve the problems and issues to deal with vague fuzzy data or in a fuzzy form through which the fuzzy systems are allowed to control certain random levels without causing damage and endangering performance to the system. Fuzzy neural system is one of the most tangible and most successful aspects of this effort. Linking and fuzzy neural system integration could be performed in two ways: a fuzzy system augmented by a neural network to promote some of its features such as speed, flexibility and the ability to adapt known as neural-fuzzy system NFS or ANFIS an equipped neural network with the ability to handle and control fuzzy information (fuzzy neural network). The design of adaptive neural-fuzzy system is orchestrated to induce the fuzzy reasoning processes where network connections are consistent with the parameters of fuzzy reasoning. Using fuzzy clustering for color image segmentation [2] is one of the methods of segmentation

* Corresponding author: john.bush@neu.edu.tr

based on pixels, where the fuzzy system determines each pixel belongs to which category. Therefore, the purpose is to create a fuzzy system that can classify more colors. To do so, an expert is required to sets the rules and membership functions according to the training data which is very time-consuming and cumbersome and the final rules might not be the best ones. Therefore an automatic method is required to create fuzzy rules and membership functions according to the training data. Many techniques have been developed for this purpose. In this article ANFIS method has been devised to produce membership functions and fuzzy rules automatically. The main advantage which provides explicit meaning for skin models is their simple decision making rules and integrating them with the main advantages of nonparametric skin models that is; having lower decision making time for training and clustering. In order to perform this method a fuzzy model for detecting skin in color images is presented.

2 Related works

The three primary colors corresponding to RGB are: red, green and blue, respectively. The RGB color components are normalized to reduce the dependence on lighting so that, sum of the normalized components is unity i.e r + g + b = 1. The third component does not hold any significant information since the sum of these components is 1 and is normally dropped for dimensionality reduction purposes. Under certain assumptions, it is observed that the differences the skin-color clusters in RGB space have relatively lower variance than the corresponding clusters in RGB therefore, are demonstrated to be useful for detection and modeling of skin color [3, 4]. Regarding the advantages mentioned above, a popular choice for detection of skin is the RGB and has been utilized by Brown et al. [5], Oliver et al. [6], Caetano and Barone [7], Bergasa et al. [8], Kim et al. [9], Sebe et al. [10], Soriano et al. [11], Schwerdt and Crowley [12], Wang and Sung [13], Storring et al. [14], Yang and Ahuja [3], Iraji [15] and Yang et al. [4]. Commission Internationale de l'Eclairage (CIE) system describes a luminance component Y as color, and other two components X and Z. The values of CIE–XYZ were constructed from psychophysical experiments and correspond to the color matching characteristics of human visual system. It is confirmed that Chen and Chiang [16], Brown et al. [5] and Wu et al. [17] used this color space. As described by Lior Shamir [18], a human perception based approach is linked to segmentation of pixel color. The H, S and V components defines fuzzy sets of the HSV color space and generate a model of fuzzy logic that aims to follow the classification of color based on human intuition. Simple modification of the classification by knowledge-driven model based on needs of a specific application and the efficiency of the algorithm in terms of the computational complexity makes the proposed method suitable for applications where efficiency is a primary issue. M. Hamidi and A. Borji [19] using fuzzy logic, have proposed a new method for color image segmentation where they automatically produce a system for classification of color and segmentation of image with minimum error rate and least number of rules. To find optimal membership functions and fuzzy rules, a comprehensive learning particle swarm optimization technique is used as it discourages premature convergence. Compared to other approaches such as ANFIS, less computational load is needed when using this method. Variety of large train data set makes the proposed algorithm invariant to noise illumination.

Rajen Bhatt, Abhinav Dhall, IIT Delhi first compiled this dataset in 2009 [20] where they published a paper titled "efficient skin region segmentation using low complexity fuzzy decision tree model" in the Indian IEEE conference known as INDICO1. Same researchers published a paper titled "adaptive digital makeup" in the same year in the international symposium on visual computing (ISVC). Leonid Sigal et al. [21] in real time systems and video conferences put forward a novel method for skin segmentation. Despite the great variety of skin types this segmentation enjoyed a fairly reliable performance. Even in brightness changes, this approach was effective. Furthermore, the Markov's method is used in this study [22]. An algorithm for segmentation of skin region in color images using edge and color information was presented by Bouzerdoum. Historically, the skin color regions were used for detection of human skin color using Bayesian model [23]. Considering high processing speed, Rodrigo Verschae et al. performed segmentation of skin regions by examining the closest pixels [24]. A new method to solving the segmentation of skin problem was designed by Junwei Han et al. [25]. At first, given a gesture of video sequences, a general skin model was developed. The data were automatically gathered from several frames. Conclusively, an SVM classifier was utilized for the detection of skin pixel on active learning basis [25]. In [26], Mohammad Shoyaib et al. performed detection of skin utilizing color distance map. Moreover, a novel recognition technique of the face was introduced; a hybrid of main components analysis and skin color segmentation known as SCS-PCA [27]. In [28], K. K. Bhoyar et al. used a classification method known as novel neural network symmetric to detect pixels of skin from non-skin ones in color images. Methods of data mining such as k-means were as well utilized for detection of skin and worthwhile results were obtained as demonstrated in [29]. In this study [29], Hamid A. Jalab using a cluster pixel model designed segmentation of skin under specifiable environmental conditions. This proposed model can overcome the changes in complex backgrounds and sensitivity to brightness conditions [30]. In [31], P. Kakumanu et al. presented modeling and detection of skin and this is recommended for further study. In conclusion, associated with the current issue, the most outstanding results is obtained using fuzzy system combined with support vector machine (FS-FCSVM) [32].

3 Skin detection using ANFIS

Adaptive Neuro Fuzzy Inference Systems (ANFISs) as explored in this paper is a rule and knowledge based system. ANFIS focal point is the knowledge base that is made up of the if-then fuzzy rules. Convincingly, if-then expression is simply the if-then fuzzy rules which are specified by a continuous membership function [33, 34]. In this paper, we explored ANFIS algorithm. An algorithm is the basic component of any intelligent system. An algorithm processes and generates knowledge from data [35, 36]. Creating systematic algorithmic models encompasses many steps. The first step is data preprocessing which is an important step in data mining; it filters and makes data ready for operations. Analyzing unfiltered data can generate inappropriate model or misleading results. Hence, the representation and quality of data is first and foremost before further analysis and classification. The second step as featured in this paper is feeding the processed data onto the classifier; ANFIS models.

This paper proposed an intelligent ANFIS for detection of skin from nonskin. This proposed algorithm is capable of detecting skin by taking mixture of RGB color space of both skin and nonskin pixels. The design stages of ANFIS are studied and implemented in this section. Beginning with the dataset, the data matrix with dimensions utilized by the algorithm is of 4*245057 (four columns and 245057 rows) where the first three columns represents R, G, and B i.e X1, X2, and X3 features and the fourth column represents the class labels; 1 for skin and 2 for non-skin. Here, detection of skin from nonskin is carried out by introducing the three R, G, and B components onto the model. Considering the extremely large number of rows, the holdout cross validation technique was explored to randomly sample the dataset into training and test sets, to minimize the biasness of the model. For high performance accuracy to be obtained, the dataset were normalized/rescaled to the range of [0, 1]. For higher system flexibility, MATLAB R2015a was used for designing and implementing the ANFISs. Figure 1 below demonstrates the framework flowchart.

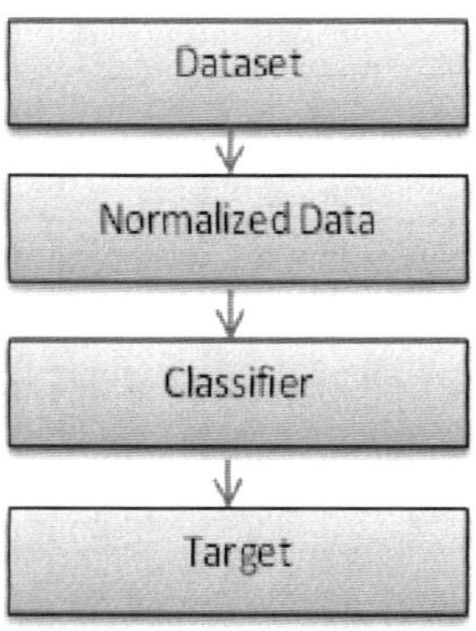

Fig. 1. Flowchart of the framework

The ANFIS morphology used in this study is made up of five layers of nodes from which the first and the fourth layers composed of the adaptive nodes having defuzzification and fuzzification incorporated, while the second, third and fifth layers comprise of fixed nodes with rule, neuron summation and normalization incorporated. Moreover, in subsequent iterations, the adaptive nodes are linked with their respective parameters constantly updating each node, and all other parameters are devoid by the fixed nodes. Architecture of the ANFIS was built on two fuzzy if-then rules using the first order Sugeno model as also demonstrated in [37-39].

First Rule: If (x is A_1) and (y is B_1) then (f_1 = p1x + q1y + r1)

Second Rule: If (x is A_2) and (y is B_2) then (f_2 = p2x + q2y + r2)

From the above two fuzzy rules, X and Y are predefined membership functions, B_i and A_i are membership values, q_i, p_i, and r_i are consequent parameters constantly updated in the forward pass in the learning algorithm and finally, f_i are the outputs. The implemented ANFIS architecture displaying those two fuzzy rules is depicted in Figure 2 where a square represents an adaptive node and a circle represents a fixed node.

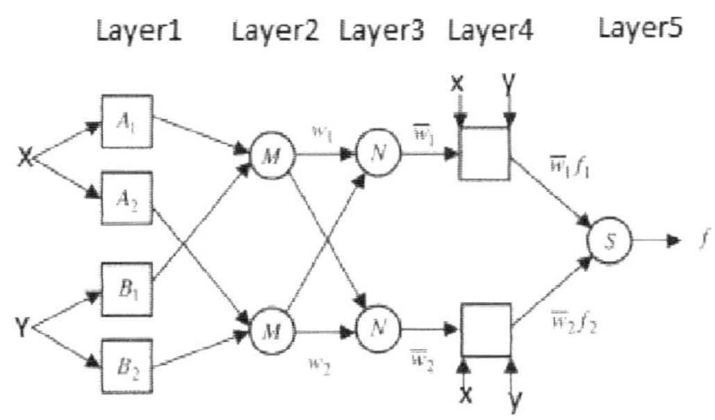

Fig. 2. Architecture of the ANFIS

From the above ANFIS structure,

- Layer 1 represents the fuzzification layer where every node i in the layer is an adaptive node. Outputs of this layer represents the fuzzy membership grade of the inputs given by:

$$O_i^1 = \mu A_i(x), for\ i = 1, 2 \quad (1)$$

$$O_i^1 \mu B_{i-2}(y), for\ i = 3, 4 \quad (2)$$

Where y and x are the inputs to node i, A is a linguistic label; small, large and $\mu_{Ai}(x)$, $\mu_{Bi-2}(y)$ assumes all fuzzy membership function.
$\mu_{Ai}(x)$ was chosen to be bell-shaped with maximum equal to 1 and minimum equal to 0, so that:

$$\mu A_i(x) = {}^1/_1 + \left\{(x - \frac{ci}{a_i})^2\right\} b \quad (3)$$

From equation 3, a_i, b_i and c_i represent the membership function parameters. Here, parameters are referred to as premise parameters.

- Layer 2 represents rule layer having a fixed node whose output is the product of all the incoming signals. Layer 2 output is depicted thus:

$$O_i^2 = w_i = \mu A_i(x)\ \mu B_i\ (y) i = 1, 2 \quad (4)$$

- Layer 3 represents the normalization layer having a fixed node represented by circle node:

$$O_i^3 = w_i = \frac{w_i}{(w1 + w2)i} = 1, 2 \quad (5)$$

- Layer 4 represents the defuzzification layer having an adaptive node. Each node's output in this layer is simply a first order polynomial and the product of the normalized firing strength as shown in equation 6:

$$O_i^4 = w_i f_i = w_i (p_i x + q_i y + r_i) i = 1, 2 \quad (6)$$

- Layer 5 represents the Summation neuron having a fixed node computing the overall output as the summation of all incoming signals as demonstrated below:

$$O_i^5 = \sum 2\ w_i f_i = \sum 2\ i = 1 w_i f_i / (w1 + w2) (7)$$

This Hybrid technique; neuro-fuzzy brings learning capabilities of neural networks to fuzzy inference systems. Where the membership functions of the Sugeno-type fuzzy inference system are fine-tuned by the learning algorithm using the training input and output data, checking input and output data; to minimizing over-fitting and finally a back propagation algorithm in combination with a least squares type of method. The proposed ANFIS classifier is presented as dermatologist's diagnostic tool in analyzing human skin as well as identifying skin from nonskin. It uses a hybrid approach of adaptive neuro-fuzzy inference system comprising two intelligent techniques; neural network (NN) and fuzzy inference system (FIS) to obtaining good reasoning in quantity and quality. That is, it is programed using NN and Fuzzy reasoning calculations.

In this paper, the classification objective is to classify skin from nonskin in a given dataset. The feature vectors extracted from RGB colour space were used as the input to the ANFIS model. Afterwards, a smaller number of iteration steps were converged using ANFIS; the hybrid learning algorithm. As depicted in figure 2, implementation of the ANFIS networks has a total of 8 fuzzy rules for experiment 1, 27 fuzzy rules for experiment 2 and 1 target for the two experiments. The ANFIS classifiers used Gbell membership functions (MFs). Choice of membership function selection is basically dependent on the user's application since there is no rule/regulation in choosing them. Here, the general ideal would be to use minimum training parameters to generate the best smallest error measure. Accepted error measure virtually determines the number of epochs fixed by the user. In this paper, 100 epochs was fixed for Experiment 1 and 2. First and second experiments resulted to 89.40% and 90.10% respectively. Results of the proposed experimental models are extensively depicted in table 1.

Table 1. Experimental results

Learning Parameters	Experiment 1	Experiment 2
Number of extracted feature vectors (model input)	3	3
Epochs	100	100
Number of fuzzy rules	8	27
RMSE	0.38	0.34
Recognition rate	89.40%	90.10%

Table 1 above demonstrates the experimental results. In both experiments, every other parameter was left constant except number of fuzzy rules. In Experiment 2, the number of fuzzy rules was increased from 8 to 27 and this raised the recognition rate from 89.40% to 90.10%. The result of these experiments generates a conventional rule; increase in the number of fuzzy rules increases the performance of the ANFIS technique. Figure 3 depict error tolerance curve for Experiment 2 (model having the best performance accuracy).

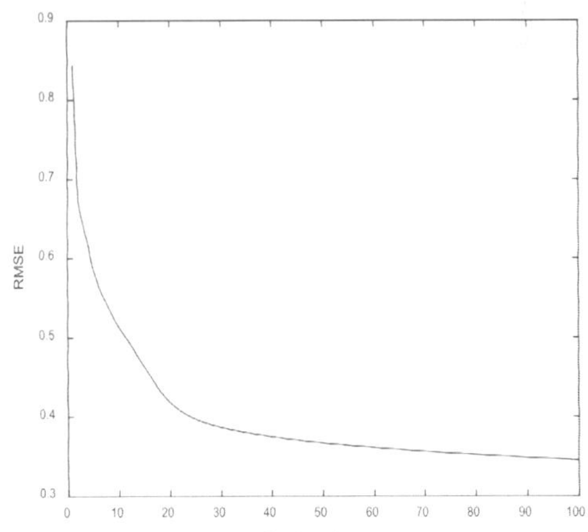

Fig. 5. RMSE learning curve

4 Conclusion and future thought

This study presents an Adaptive Neuro Fuzzy Inference System with very high potentials to detecting human skin from nonskin. Our proposed algorithm is capable of simulating human knowledge and experience from complex decision makings using approximate information and in environments with uncertainty. The explored approach operates in accordance with particular characteristics for detecting human skin from nonskin. Considering the input data, detection is categorized as very low, low, median, and high. Regarding the obtained results from the two experiments, second experiment with 90.10% accuracy outperformed first experiment having 89.40% accuracy. A dermatologist at the prestigious Near East University Hospital-Cyprus considered this result to be worthwhile. Interestingly, between the two experiments; Experiment 1 having low fuzzy rules and Experiment 2 with high fuzzy rules, it was recorded that Experiment 1 having low fuzzy rules was outperformed by experiment 2 having high number of fuzzy rules. The experimental results show that the high number of fuzzy rules used in Experiment 2 actually boosted the performance of the system by increasing the performance accuracy by 0.7%. Observations from this experimental results proved that it is totally advisable to use moderately high number of fuzzy rules in ANFIS computation since further increase of the number of rules would result to reasonably high classification and performance accuracy. Furthermore, we are of no claim that our proposed system has the best performance ever. To improve on the performance, future contributions to this problem would feature the repetition of the experiment using other machine learning techniques such as co-adaptive neuro-fuzzy inference system (CANFIS), extreme learning machines (ELMs), deep learning and support vector machines (SVMs) to obtaining more optimal results.

References

1. A. Iraji, M. Saber, and A. Tosinia, Skin color segmentation in YCBCR color space with adaptive fuzzy neural network (Anfis), Int'l J. Img. Graphs. Sig. Proc. 4.4: **35** (2012)
2. B. Akshay, S. Srivastava, A. Agarwal, Face detection using fuzzy logic and skin color segmentation in images, *Emerging Trends in Engineering and Technology (ICETET), 3rd International Conference*, IEEE, (2010)
3. M. Yang, N. Ahuja, Gaussian mixture model for human skin color and its application in image and video databases, SPIE: *Conference on Storage and Retrieval for Image and Video Databases*, 3656:458–**66**, (1999)
4. J. Yang, W. Lu, A. Waibel, Skin-color modeling and adaptation, ACCV98; (1998)
5. D. Brown, I. Craw, J. Lewthwaite, A SOM based approach to skin detection with application in real time systems, BMVC01; (2001)
6. N. Oliver, A. Pentland, F. Berard, Lafter: Lips and face real time tracker," CVPR97. (1997)
7. T. Caetano, D. Barone, A probabilistic model for the human skin-color, ICIAP01; p. 279–**83**, (2001)
8. L. Bergasa, M. Mazo, A. Gardel, M. Sotelo, L. Boquete, Unsupervised and adaptive Gaussian skin-color model, *Image and Vision Computing*, **18**,12:987–1003, (2000)
9. S. Kim, N. Kim, S. Ahn, H. Kim, Object oriented face detection using range and color information, AFGR**98**, (1998)
10. N. Sebe, T. Cohen, T. Huang, T. Gevers, Skin detection, a Bayesian network approach, ICPR**04**, (2004)
11. M. Soriano, J. MartinKauppi, S. Huovinen, M. Laaksonen, Adaptive skin color modeling using the skin locus for selecting training pixels, *Pattern Recognition Letters*, **36**,3:681–90 (2003)
12. K. Schwerdt, J. Crowely, Robust face tracking using color, AFGR00, (2000)
13. J. Wang, E. Sung, Frontal-view face detection and facial feature extraction using color and morphological operations, *Pattern Recognition Letters*, **20**:1053–68, (1999)
14. M. Stoerring, T. Koeka, H. Anderson, E. Granum, Tracking regions of human skin through illumination changes, *Pattern Recognition Letters*, **24**,11 (2003)
15. Iraji, M. Saber, A. Yavari, Skin color segmentation in fuzzy ycbcr color space with the mamdani inference, US. J. Sc. Res. **7**:131–7 (2011)
16. C. Chen, and S. Chiang, *Detection of human faces in colour images*, IEEE *Proceedings of Vision Image and Signal Processing*, **144**,6:384–8 (1997)
17. H. Wu, Q. Chen, M. Yachida, Face detection from color images using a fuzzy pattern matching method, IEEE *Transactions on Pattern Analysis and Machine Intelligence*, **21**, 6:557–63 (1999)
18. L. Shamir, Human perception-based color segmentation using fuzzy logic, *International Conference on Image Processing, Computer Vision and Pattern Recognition*, IPCV , **2**: 496-505, (2006)
19. A. Borji, M. Hamidi, Evolving a fuzzy rule base for image segmentation, *Proceedings of world academy of science, engineering and technology*, **22**:4-9, (2007)
20. B. Rajen Bhatt, G. Sharma, A. Dhall, S. Chaudhury, Efficient skin region segmentation using low complexity fuzzy decision tree model, IEEE-INDICON, 1-4, (2009)
21. A. Dhall, G. Sharma, R. Bhatt, Adaptive digital makeup, *Proceedings of International Symposium on Visual Computing,* ISVC, 728–36 (2009)
22. L. Sigal, S. Sclaroff, V. Athitsos, Estimation and prediction of evolving color distributions for skin segmentation under varying, IEEE *Conference on Computer Vision and Pattern Recognition,* (2000)
23. S. Phung, A. Bouzerdoum, D. Chai, Skin segmentation using color and edge information, IEEE, (2003)
24. J. Ruiz-del-Solar, R. Verschae, Robust skin segmentation using neighborhood information, IEEE, (2003)
25. J. Han, G. Award, A. Sutherland, H. Wu, Automatic skin segmentation for gesture recognition combining region and support vector machine active learning, *Proceedings 7th International Conference on Automatic Face and Gesture Recognition,*IEEE; (2006)
26. M. Abdullah-Al-Wadud, M. Shoyaib, O. Chae, A skin detection approach based on color distance map, EURASIP, *Advances in Signal Processing*, (2008)
27. L. Lang, G. Weiwei, The face detection algorithm combined skin color segmentation and PCA, IEEE, (2009)
28. K. Bhoyar, O. Kakde, Skin color detection model using neural networks and its performance evaluation, JCS, **6**, 9:963–8, (2010)
29. P. Ng, P. Chi-Man, Skin color segmentation by texture feature extraction and k-mean clustering, IEEE, *Third International Conference on*

Computational Intelligence, Communication Systems and Networks, (2011)

30. S. Naji, R. Zainuddin, H. Jalab, Skin segmentation based on multi pixel color clustering models, Elsevier Inc, (2012)
31. P. Kakumanu, S. Makrogiannis, N. Bourbakis, A survey of skin-color modeling and detection methods, *Pattern Recognition Letters*, **40**:1106–22 (2007)
32. Juang, Chia-Feng, C. Shih-Hsuan, S. Shen-Jie, Fuzzy system learned through fuzzy clustering and support vector machine for human skin color segmentation, *Systems, Man and Cybernetics, Part A: Systems and Humans*, IEEE, **37,** 6:1077–87 ()2007
33. A. Rahib, A. Nurullah, E. Aytac, I. Günsel, A. Çağman, Brain-Computer Interface for Control of Wheelchair Using Fuzzy Neural Networks, BioMed res. Int'l, (2016)
34. A. Rahib, H. Vasif, Abiyev, C. Ardil, Electricity consumption prediction model using neuro-fuzzy system, IJCEACIE, **3**, 12, (2009)
35. I. John Bush, K. Dimililer, Static and Dynamic Pedestrian Detection Algorithm for Visual Based Driver Assistive System, ITM Web of Conf. **9**, 03002. EDP S. (2017)
36. A. Helwan, D. Uzun Ozsahin, A. Rahib, I. John Bush, One-Year Survival Prediction of Myocardial Infarction, IJACSA, **8**, 6, (2017)
37. S. Abdulkadir, An expert system based on linear discriminant analysis and adaptive neuro-fuzzy inference system to diagnosis heart valve diseases, Exp. Syst. App. (2008)
38. G. Zhao, C. Peng, W. Xiting, Intelligent control for AMT based on driver's intention and ANFIS decision-making, *World Congress on Intelligent Control and Automation*, (2008)
39. D. Anupam, J. Maiti, R. Banerjee, Process control strategies for a steel making furnace using ANN with Bayesian regularization and ANFIS, Exp. Syst. App. (2009)

TECHNOLOGICAL FORMS DEVELOPE IN DIFFERENT MORPHOLOGICAL FIELDS

Delia Garleanu[1,*], *Delicia* Arsene[1], *Gabriel* Garleanu[1], and *Claudia* Borda[1]

[1]UPB, IMST Faculty, TMS Department, 060042 Splaiul Independentei 313, Bucharest, Romania.

Abstract. The present work is takes in consideration the experimental study of the influence of some morphological fields regarding the development of the complex technological forms. The goal that we have set in this paper was to bring in new points of view on technology morphologies arising in technological processes, whose goal is to generate surfaces.The development of the technological shapes could be controlled and oriented from the quality and quantity point of view, by using the morphological fields.. *We refer here in particular to complex forms, not the Euclidean upon which adequate studies have been done already.* Among the illustrations of the experimental parameters of forming events, electrochemical deposit is regarded as a paradigm for theoretical studies of diffusion limited aggregation. In fact, by changing the concentration of metal ions, the cathodic potential and the morphological field you can explore different morphologies, such as: dense radial aggregates, dendridic patterns and fractal aggregates. This topic has been treated and shown in the paper. The box-counting algorithm was used in calculating the fractal dimension of the deposited agregates.

1 Introduction

In order to bring some extra objectivity in evaluating the shapes, we have done an experiment and study regarding the Copper dendrite growth, by electrolytic deposit in a CuSO4 solution, under the presence of some Euclidian morphological fields.

In the same time, we have done a selection of the resulted structures, that were considered relevant from the auto-similarity point of view, in order to attach fractal like interpretation mathematical procedure.

As in the field, virtually no research has been done, we were required to develop investigative techniques specific to consider estimators of complex shapes and their quantitative evaluation.

2 Experimental context

We tried to achieve, through the experiments we've carried out, some electrodeposited geometries, in wide variety, in order to be able to investigate, as well as possible, the clusters shape. That's why we used different experimental conditions. We present here under the parameters that we controlled in the process.

2.1 Operation Stand

Electrolysis cell is connected to the adjustable voltage source,a current transducer is interposed between the cathode and the source mass (2Ω resistor).

Electrolysis cell anode is attached via an electrical insulator, to the pan balance. An electronic opto-coupler device is attached to the balance and a shutter is fixed to its needle. Thus the change of mass is converted by the opto-coupler, by the deflection of the needle balance into an electrical signal through the amplifier for the opto-coupler.

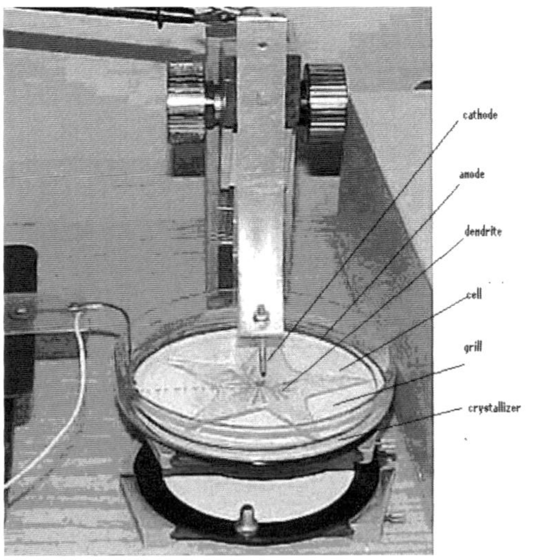

Fig. 1. Operation stand.

* Corresponding author: delia.garleanu@upb.ro

Electrical signals for the effects of mass and current are sent to the data aquisition card from the computer, thus it can be viewed as built with LabVIEW as well as on an oscilloscope and a multimeter.

Further, in detail, is presented the electrolytic deposited cell.

It consists of:

- Crystallizer;
- Cu Anode ring;
- Cathode brass cylinder, 1.5 mm in diameter and 3mm length;
- CuSO4 solution;

To have the prospect of dendritic image formed by scanning, cathode is mounted on a glass plate with different geometric shapes.

Under crystallizer it was placed a concentric grid while tracking the development of the cluster radial electrodeposition.

2.2. Introducing morphological fields

The electrolyte type: Copper sulphite and the quality of the soluble anode, that was electrolytic copper.The thickness of the electrolyte layer: it was limited at 0,5 – 1 mm in order to prevent the growth of the electro-deposited cluster but only in 2D and not in 3D dimension.

The Copper sulphite solution's concentration: there have been carried out several procedures, with different concentrations of the solution, from 5% to 30%. The 30% concentration was the saturation one. Within 5% to 10% concentration the process was very slow and the deposited material was insignificant. For the 15% concentrations the fractals' geometries were highly ramified but that was dependent on the current parameters, tension and intensity. For the higher concentrations, bigger than 15%, the dendrite patterns are more dense, the ramification capacity decreases and the main growth directions are most visible ones.

The tension and intensity of the electric power: from this point of view there have been observed two cluster categories: those which have grown under constant tension and those which have grown under constant intensity.

The clusters that grown under constant tension presented ramified geometries, between the dendritically and fractal patterns, depending on the concentration of the electrolyte solution:

- for the concentrations between 15% and 20% there take place the ramified growth with self-scaling capacity.

- for the concentrations over 20% up to saturation, there begin to show up, step by step, dens morphologies, with faint or with no ramifications.

There have been used different tensions for the current but the most interesting results have been obtained for U=10V. The intensity varied from 0,18A to 1A. The clusters that grew under constant intensity show a limitation of the growth diameter, after the cathode potential was decreased to 6 V. From here on the experiment reveals that the deposited material grows up to the limits of the aggregate, extending its surface a lot, in the volume that the liquid level allows. In this case it was realised a 3D fractal aggregate that was limited at the cluster's margins.

The intensity of the current had the limits between 0,2A and 0,8A, but the most spectacular results were obtained between 0,4A and 0,6A; and this happened because for 0.4 A the growth was very slow and did not go over 30 – 35 mm diameter and for intensities bigger than 0,6A we should have grown the tension over 15V which would have conduct to boiling phenomena around the cathode.

We present in figure 2 some of the results of these experiments.

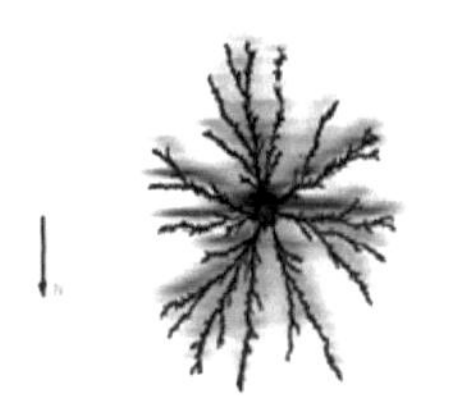

a) Ramified cluster grown on the rectangular support

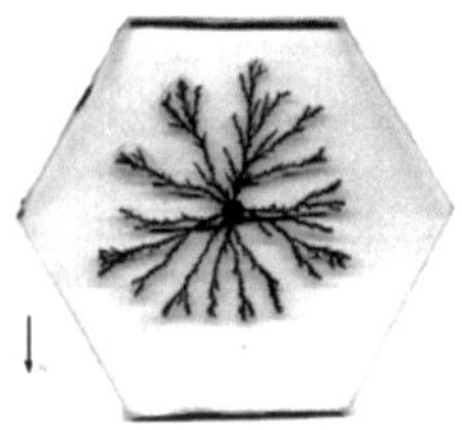

b) Ramified cluster grown on the hexagonal support

c) Ramified cluster grown on the stellar pentagon support

d) Ramified cluster grown on the isosceles triangle support

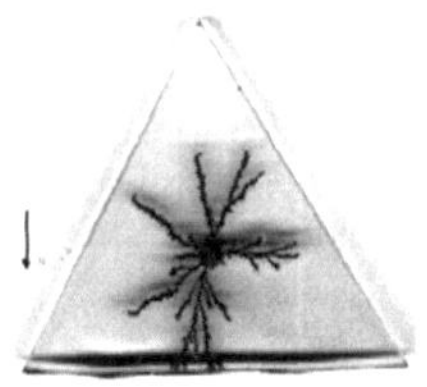

e) Ramified cluster grown on the equilateral triangle support

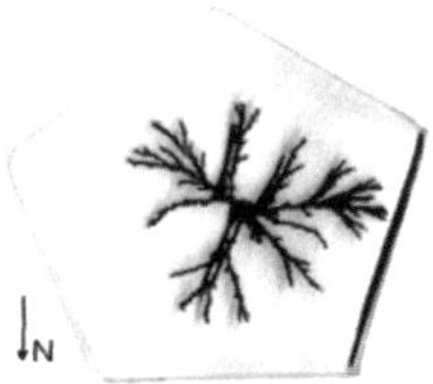

f) Ramified cluster grown on the regular pentagons support

Fig. 2. Ramified clusters grown in some morphological fields in electrolyte solution with the weak concentrations (between 15% and 20%)

The geometry of the morphological fields: it was introduced through a glass support, on which we fixed the cathode and where the clusters were developing.

https://doi.org/10.1051/itmconf/20181602005

In these conditions, the image capture, used for scanning, was eased a lot. There were used different shapes for the glasses supports, such as circles, with different diameters, regular pentagons, stellar pentagons, hexagons, triangle, square forms.

The influence on the morphological fields was more obvious for the circles, stellar pentagon and hexagon.

For the circle shape supports, the cluster had a constant growth speed and approximately equal on each brunch; all the brunches reached the same diameters in the same time. As for the stellar pentagon there have been observed a bigger growth in favour of those directions not oriented to the corners and even though the deposited material was made after radial patterns, these directions were more favourite.The clusters in the hexagonal supports showed regular geometries, with approximately equal angles between the main ramifications and similar growth, as number and angles, on each brunch. The other morphological fields did not have such a strong influence on the clusters geometry.

In the figure 2 we show the clusters with some geometries, grown in the constant intensity, that were obtained by electrodepositing: strongly ramified, with self-scaling capacity, weakly ramified or not ramified at all, radial.

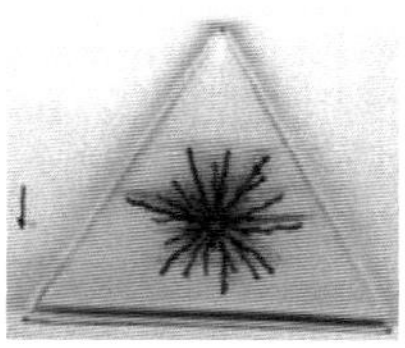

a) Unramified radial cluster grown in the weak concentration solution

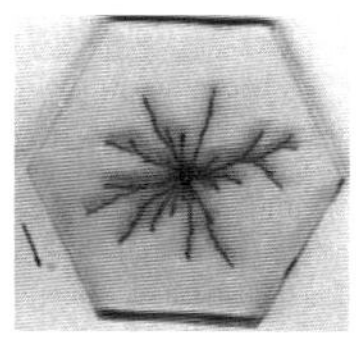

b) A little ramified cluster grown in the weak concentration solution

c) Radial cluster limited in diameter, grown in the weak concentration solution

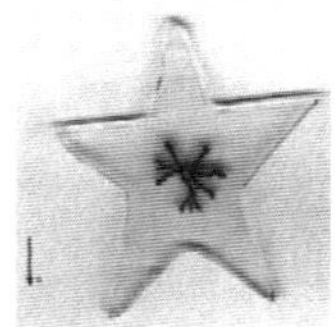

d) Radial cluster limited in diameter, grown in the weak concentration solution

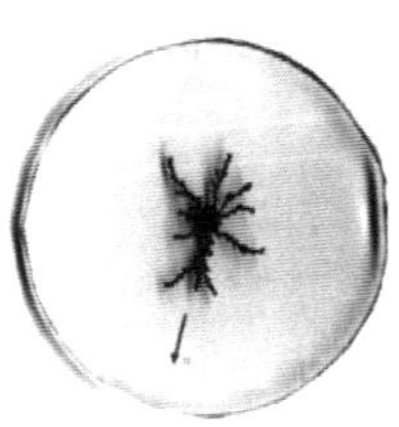

e) A little ramified cluster grown in the high concentration solution

Fig. 2. Clusters with some geometries, grown in the constant intensity.

There have been tried a certain positioning after the cardinal points and the moment of the day when the experiment took place but there was no connection between these parameters and the shape or size of the electro-deposited shape.

It is important to notice that the study of the self similar fractal aggregates shows that these morphologies are to be found as product of other processes and phenomena In this respect, there could be applied a fractal mathematical algorithm, that already exists and could be used in order to study these objects.

There is to be mentioned that the angles between the aggregates ramifications are pentagonal geometries with 36^0 main values. As for the ramification value, this was level 3, as for an object is considered to be fractal if it is capable for scaling at least on 3 levels. So, in this study we didn't give any attention to the non-ramified geometries or to those weakly ramified and not even the radial ones, because none of them had self-similarity for more than 2 levels scaling.

2.3 Generalized fractal dimensions

In the context of this paper, the generalized fractal dimensions will be used to quantify the self similarity properties of numerical fractal aggregates and experimental ones.

When investigating scaling properties of fractal geometry aggregate measure μ is the mass of the object.

To define a square with a uniform grid ε, size (Fig. 4.1). Let I denote by μ_I proportion of the total mass of the object that is contained inside the box "and" for an "i": for an empty box ε does not exist.

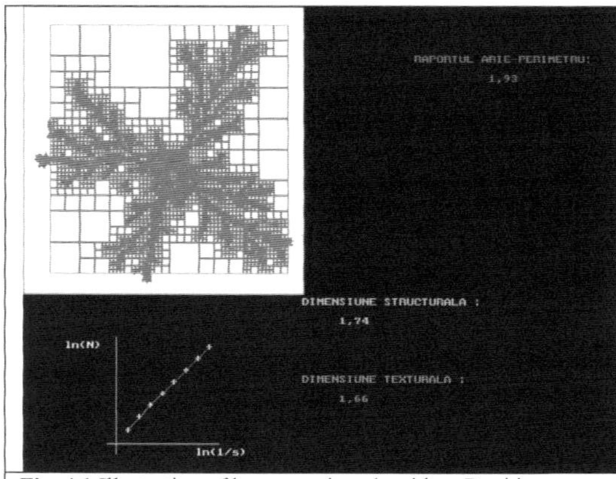

Fig. 4.1. Illustration of box-counting algorithm. Partition function zq (ε), is calculated using a grid of square boxes ε. side. D_q dimensions are estimated by averaging the random positions on the grid in accordance with the unit.

As we know, one of the essential characteristics of fractal objects is that they manage to pack in a relatively small zone very high perimeters, in case of two-dimensional structures, or in relatively small volumes large areas for 3D structures. If we look at 2D and to define the terms "relatively small" and "relatively large" we say that the Euclidean perimeter which delineate a surface area increases theoretically unlimited when it is a fractal object, depending on the unit size that measured it. So we can say that if:

$l \to 0$ then

$p \to \infty$ and

$A \cong ct.$

where: l is the length unit which made the perimeter measurement; p – the fractal object perimeter; A – the fractal object area.

These unlike the Euclidean figures where the perimeter size does not depend on the length unit of which the measurement is made. So we thought appropriate that in addition to other morphological investigation techniques presented to introduce the area-perimeter relation, we can give us a measure of the complexity of object edges, a measure of the packing degree; that's why we will call it, the wrapping dimension.

2.4 The work algorithm

We call the packing dimension the perimeter log slope graph function of log area length unit which we measure it.

Both the generated articles by iterative functions and for the natural objects (the Cu electrolytic depositions) is measured with perimeter different lengths. It plots log (Perimeter) / log (measurement unit) and the obtained graph slope is calculated by the interpolation function. This is the packing dimension. It characterizes the subject from the global point of view.

The fractal dimensions obtained for the deposited clusters varied between 0, 8 and 1.98 as follows:

- for simple Euclidean square shapes, circle and weak consonants of the solution the fractal dimensions were in the beginning of the size range
- for Euclidean forms with multiple sides, the fractal dimensions were increased with the complexity of the geometry and the concentration of the solution

Any object is an information carrier and the higher the delimiter surface is, the greater the storage capacity of the information and the interfacing with the environment in which it is sunk.

Considering the above, we can introduce a controlled development of such information-bearing objects, through the variation of the morphological fields and the parameters that condition the formation of clusters.

Also, we have introduced an algorithm for calculating the packing size for the crossing into concrete of the complex structures characteristics.

The packaging size provides a clear relationship between the volume of the object and the surface that encompasses it.

The lower the volume and the larger the packaging surface, the greater the storage capacity and the more intense the exchange of information are..

3 Conclusions

The study of the self similar fractal agregates shows that these morphologies are to be found as product of other processes and phenomena there could be applied a fractal mathematical algorithm, that already exists and could be used in order to study different objects.

The development of the technological shapes could be controlled and oriented from the quality and quantity point of view, by using the morphological fields.

The introduction of some characteristic features of the surface/perimeter ratio contributes to the quantification of complex structures.

In this respect, technologies can be developed to produce objects with a high capacity for storing information in small volumes and to convey an important information flow.

We intend, in the future, to control the growth by using "growth morphological inhibitors" and "field concentrators" which should introduce morphological fields of different qualities.

References

1. D. Arsene, *Contributions regarding the technological shapes stability and their interpretation through the fractal theory*, Doctor Thesis, University Politechnica Bucharest, (1999)
2. G. Palla, P. Pollner and T.Vicsek , *Rotated multifractal network generator*, Journal of Statistical Mechanics: Theory and Experiment (2011).
3. Z. Kazim, *The Hausdorff and box dimension of fractals with disjoint projections in two dimensions,* Glasgow Math J 44:117-123 (2002).
4. Lapidus ML, van Frankenhuijsen M, *Fractal geometry and Applications* , part 1, 2 American Mathematical Society, Providence (2004).
5. Y. Heurteaux, *Dimension of mearures; the probabilistic approach,* Publ.Mat.51, 280-290, (2007).
6. Ben Avraham D., Havlin S., *Diffusion and reactions in fractals and disordered systems*, Cambridge University Press, Cambridge (2000)
7. Jaffard S, Meyer Y, Ryan RD, *Wavelets: Tools for science and technology,* SIAM, Philadelphia (2001).
8. Kenneth Falconer, *Fractal Geometry: Mathematical Foundations and Applications*, 3rd Edition, , ISBN 978-1-119-94239-9, 398pg, Jan (2014).
9. D-J Feng, KS Lau, *Multifractal formalism for self-similar measures with weak separation condition*, J. Math.Pures Appl., 92, (2009).
10. D. Moise, B. Bogdan, D, Druta, *Aldoritmi, numere si fractali,* Ed Printech, (2007).
11. A. Bunde, S. Halvin, *Fractals and disordered systems*, Springer Science, (2012).

 https://doi.org/10.1051/itmconf/20181602006

Performance comparison of model selection criteria by generated experimental data

Radoslav Mavrevski[1,*], *Peter* Milanov[2,3], *Metodi* Traykov[1], *Nevena* Pencheva[4]

[1]Department of Electrical Engineering, Electronics and Automatics, Faculty of Engineering, University Center for Advanced Bioinformatics Research, South-West University "Neofit Rilski", 66 Ivan Mihaylov Str., 2700 Blagoevgrad, Bulgaria

[2]Department of Informatics, Faculty of Mathematics and Natural Sciences, University Center for Advanced Bioinformatics Research, South-West University "Neofit Rilski", 66 Ivan Mihaylov Str., 2700 Blagoevgrad, Bulgaria

[3]Institute of Mathematics and Informatics, Bulgarian Academy of Sciences, Acad. Georgi Bonchev Str., Block 8, 1113 Sofia, Bulgaria

[4]Department of Health Cares, Faculty of Public Health, Health Cares and Sports, University Center for Functional Research in Sports and Kinesitherapy, South-West University "Neofit Rilski", 66 Ivan Mihaylov Str., 2700 Blagoevgrad, Bulgaria

Abstract. In Bioinformatics and other areas the model selection is a process of choosing a model from set of candidate models of different classes which will provide the best balance between goodness of fitting of the data and complexity of the model. There are many criteria for evaluation of mathematical models for data fitting. The main objectives of this study are: (1) to fitting artificial experimental data with different models with increasing complexity; (2) to test whether two known criteria as Akaike's information criterion (AIC) and Bayesian information criterion (BIC) can correctly identify the model, used to generate the artificial data and (3) to assess and compare empirically the performance of AIC and BIC.

1. Introduction

Model selection is a process of choosing a model from set of candidate models from different classes, which will provide the best balance between goodness of fitting of the data and complexity of the model [1, 2, 3]. There are different criteria for evaluation of competitive mathematical models for data fitting (approximation). Information criteria provide an attractive base for model selection [1, 3], [4-11]. However, little is understood about their relative performance in model selection.

This research has several specific objectives:

(1) to generate artificial experimental data by known test models;

(2) to fitting data with various models with increasing complexity;

(3) to verify if the class model used to generate the data could be correctly identified through the two commonly used criteria Akaike's information criterion (AIC) and Bayesian information criterion (BIC) and to assess and compare their performance.

2. Materials and methods

2.1 The generate of experimental data

We use the GraphPad Prism software for the artificial experimental data generating and for curve fitting. GraphPad Prism combines nonlinear regression, basic biostatistics, and scientific graphing. (http://www.graphpad.com/scientific-software/prism). To generate artificial experimental data we use class model - third order polynomial. The individual member of this class is:

$$y = 44 + 99x - 59x^2 + 8x^3 + \varepsilon \quad (1)$$

$$\varepsilon \sim Normal,\ SD = 5,$$

where ε is random error with Gaussian distribution and standard deviation (SD).

The graph of the third order polynomial:

$a + bx - cx^2 + dx^3$, *a=44, b=99, c=-59, d=8* is shown in Figure 1.

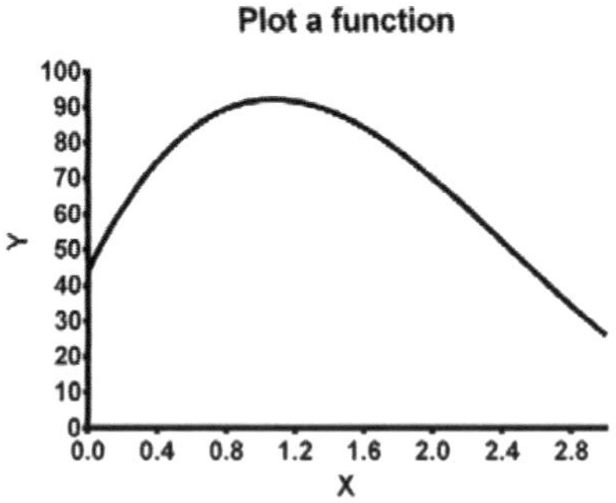

Fig. 1. The individual member of third order polynomial.

* Corresponding author: radoslav_sm@abv.bg

For our computational experiments were generated samples with different sizes - small sample (15 points), middle sample (31 points) and large sample (101 points), following the classification in [12].

2.2 Fitting experimental data

In this research we use different class models (polynomials from first to sixth order) for fitting the artificial experimental data. To find the individual "optimal" models $P^*(M_j)$ in the classes M_j, $j = 1, \ldots, 6$, we use least squares fitting in GraphPad Prism 6.0. Least squares fitting criterion is defined as follows:

$$F(a) = \sum_{i=1}^{n} (y_i - f(x_i, a_1, \ldots, a_s))^2 \qquad (2)$$

The problem is to find $a^* = (a_1^*, \ldots, a_s^*)$, such that minimizes $F(a)$.

2.3 Criteria for selection of the optimal model from different class models

2.3.1 Akaike's information criterion

One of the most commonly used criterion for model selection is AIC. The idea of AIC is to select the model that minimizes the negative likelihood penalizing by the number of parameters:

$$AIC = \begin{cases} n ln\left(\frac{RSS}{n}\right) + 2k, & \frac{n}{k} \geq 40 \\ n ln\left(\frac{RSS}{n}\right) + 2k + \frac{2k(k+1)}{n-k-1}, & \frac{n}{k} < 40, \end{cases} \qquad (3)$$

where n is the number of data points; k is the number of the fitting parameters by the regression plus one (since regression is "an estimating" of the sum-of-squares as well as the values of the parameters); RSS, or residual sum of squares, is the sum of the squares of the vertical deviations from each data point to the graph of a curve of the "optimal" fitted model.

2.3.2 Bayesian information criterion

The other most commonly used criterion BIC has the highest posterior probability. AIC and BIC criteria differ only in that the coefficient multiplies the number of parameters. In other words, the criteria differ by how strongly they penalise large models:

$$BIC = n ln\left(\frac{RSS}{n}\right) + k ln(n), \qquad (4)$$

with the same meaning of RSS, n and k, above.

In this situation, the model that minimizes BIC has the highest posterior probability. BIC penalizes the models more from AIC for increasing number of parameters. AIC does not depend directly on the sample size. In general, models chosen by BIC will be more parsimonious than those chosen by AIC.

2.3.3 Program for calculating the AIC and BIC criteria

For calculation of the criteria values of AIC and BIC according to formulas (3) and (4), we use a program "Comparing Models" developed by us in our previous research (see Figure 2), [8].

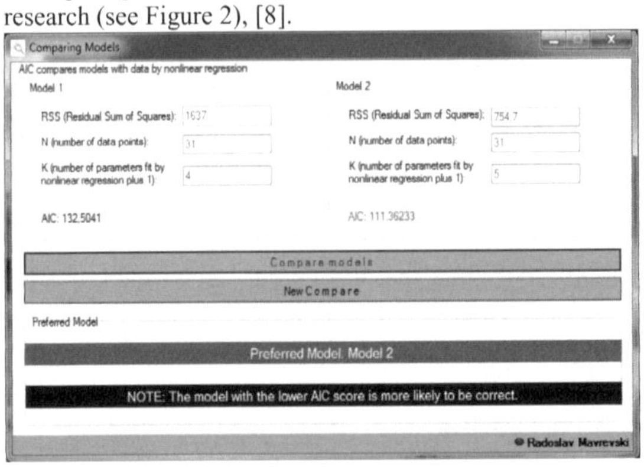

Fig. 2. Example for calculation of the AIC: dialogue box of the program "Comparing Models" for calculating AIC.

3. Results

3.1 Case 1: generate 15 points

For our first experiment we generate 15 points (small sample) in the interval from 0 to 3 with step 0.2. In this case only AIC criterion correctly identified the third order polynomial (true class model) as the optimal model, and BIC criterion chooses fifth order polynomial as the optimal model (false class model) (see Table 1).

Table 1. Result with small sample (15 points).

Polynomial class model	Number of data points	Number of parameters	AIC value	BIC value
First order	15	2	91.99	91.93
Second order		3	67.76	66.59
Third order		4	**60.00**	56.87
Fourth order		5	60.88	54.63
Fifth order		6	64.70	**53.65**
Sixth order		7	74.66	56.32

Simulated data and curves of the fitting models are shown in Figure 3.

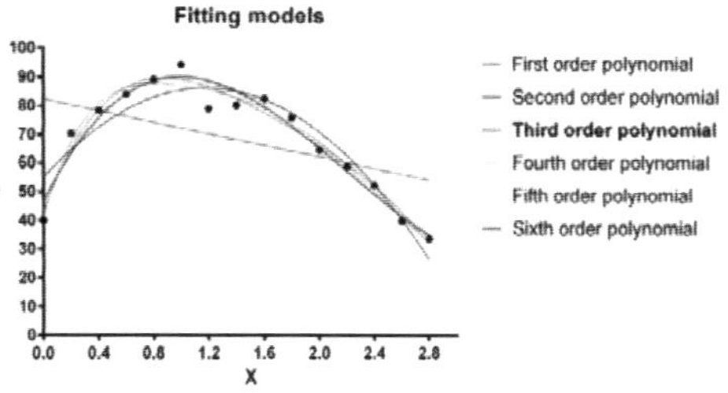

Fig. 3. Simulated data (15 points) and curves of the fitting models (six fitted polynomial curves of increasing order, from 1 (straight line) to 6).

3.2 Case 2: generate 31 points

For the second experiment we use 31 points (middle sample), that are generated in interval from 0 to 3 with step 0.2. Here, both AIC and BIC criteria correctly identified the third order polynomial (true class model) as the optimal model (see Table 2).

Table 2. Result with middle sample (31 points).

Polynomial class model	Number of data points	Number of parameters	AIC value	BIC value
First order	31	2	181.15	184.57
Second order		3	132.50	136.70
Third order		4	**111.36**	**116.13**
Fourth order		5	114.46	119.56
Fifth order		6	116.35	121.52
Sixth order		7	119.74	124.67

Simulated data and curves of the fitting models are shown in Figure 4.

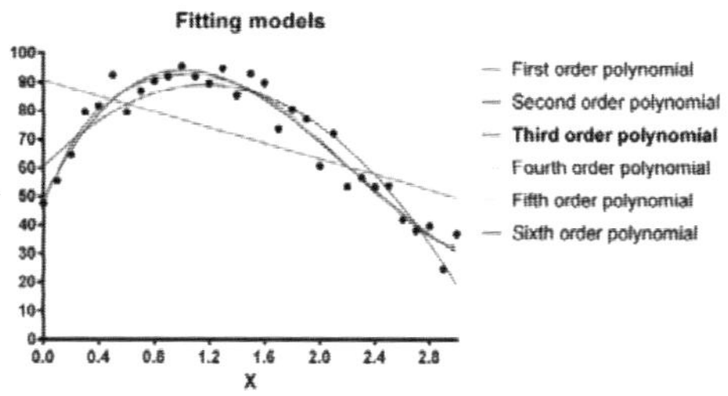

Fig. 4. Simulated data (31 points) and curves of the fitting models (six fitted polynomial curves of increasing order, from 1 (straight line) to 6).

3.3 Case 3: generate 101 points

In the last case we use 101 points (large sample) generated in interval from 0 to 3 with step 0.2. The obtained results showed that in this case only BIC criterion correctly identified the third order polynomial (true class model) as the optimal model, while AIC criterion chooses fifth order polynomial as the optimal model (false class model) (see Table 3).

Table 3. Result with large sample (101 points).

Polynomial class model	Number of data points	Number of parameters	AIC value	BIC value
First order	101	2	569.67	577.26
Second order		3	348.03	358.07
Third order		4	303.72	**316.16**
Fourth order		5	305.04	319.84
Fifth order		6	**300.47**	317.57
Sixth order		7	302.01	321.95

Simulated data and curves of the fitting models are shown in Figure 5.

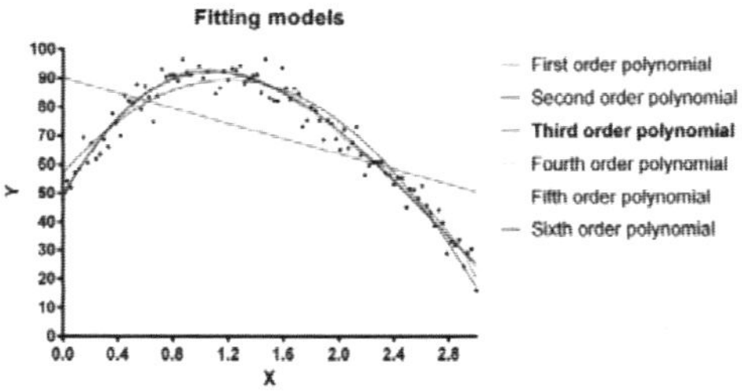

Fig. 5. Simulated data (101 points) and curves of the fitting models (six fitted polynomials curves of increasing order, from 1 (straight line) to 6).

In Figure 6 we show comparison of effectiveness of AIC and BIC in the selection of the optimal model (in all three cases) from the set of 6 class polynomials that was used for fitting the data.

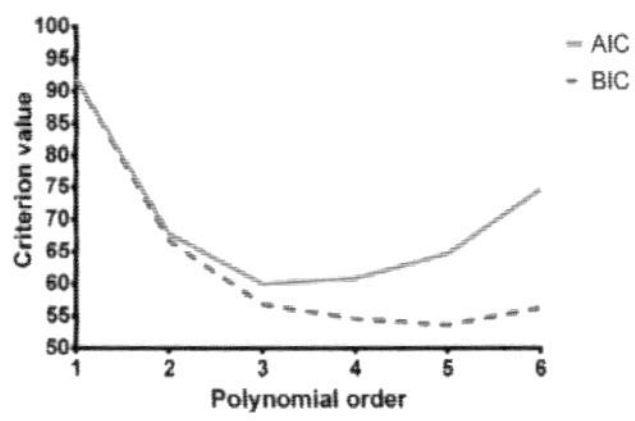

(a)

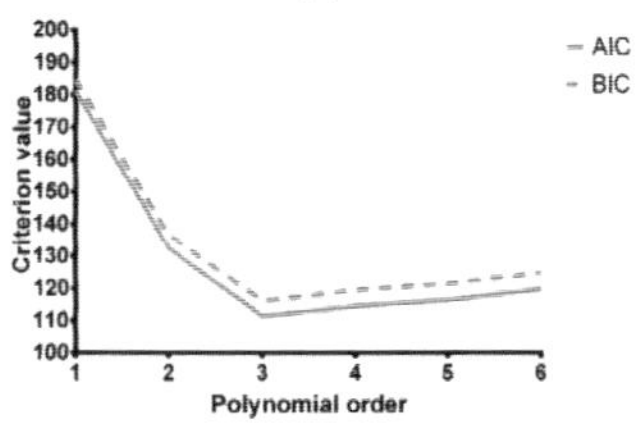

(b)

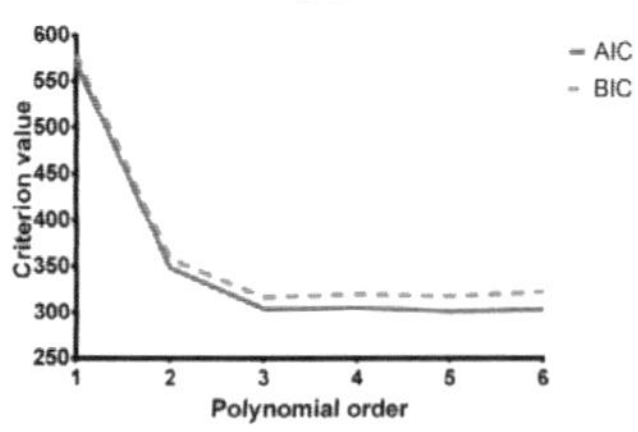

(c)

Fig. 6. Comparison of the effectiveness of AIC and BIC: (a) 15 points generated, (b) 31 points generated, (c) 101 points generated.

Figure 6(a) shows that AIC chooses the true class model, and Figure 6(c) shows that BIC chooses the true class model. In Figure 6(b) we can see that both, AIC and BIC, choose the true class model.

4. Discussion

The obtained results from the computational experiments suggested that AIC performs relatively well for small samples but is inconsistent and does not improve performance for large samples. The BIC criterion appears to perform relatively poorly for small samples but is consistent and improves its performance with increasing the sample size. This is consistent with previous studies [4, 13], which demonstrated that BIC is consistent (that is, it tends to choose the true model with a probability equal to 1) in large samples. In our experiments BIC also outperforms AIC when there is a large sample (101 data points) in identification the true class model. As a whole, the current results suggest that generally AIC should be preferred in smaller samples whilst BIC should be preferred in larger samples.

Acknowledgements

This work is partially supported by the project of the Bulgarian National Science Fund, entitled: Bioinformatics research: protein folding, docking and prediction of biological activity, NSF I02 /16 12.12.14.

References

1. H. Acquah, J. Dev. Agric. Econ. **2**, 1-6 (2010)
2. SJ. Ahn, JIPS **4**, 153-158 (2008)
3. H. Akaike, IEEE Trans. Autom. Control **19**, 716–772 (1974)
4. P. Bickel and P. Zhang, J. Am. Stat. Assoc. **87**, 90–97 (1992)
5. P. Burnham and D. Anderson, *Model Selection and Multimodel Inference 2 ed.* (Springer-Verlag, New York, 2002)
6. ID. Coope, J. Optim. Theory. Appl. **76**, 381-388 (1993).
7. B. Joseph and L. Nicole, J. Am. Statist. Assoc. **99**, 279–290 (2004)
8. R. Mavrevski, C. R. Acad. Bulg. Sci. **67**, 1345-1354 (2014).
9. T. Dzimbova, F. Sapundzhi, N. Pencheva, P. Milanov, Int. J. Bioautomation **17**, 5-16 (2013)
10. F. Sapundzhi, T. Dzimbova, N. Pencheva, P. Milanov, Bulg. Chem. Commun. **2,** 613-618 (2015)
11. F. Sapundzhi, T. Dzimbova, N. Pencheva, P. Milanov, Der Pharma Chemica **8**, 118-124 (2016)
12. A. Ghasemi and S. Zahediasl, Int J Endocrinol Metab. **10**, 486-489 (2012)
13. E. Ward, Ecol. Model. **2**, 1-10 (2008).

A new off-lattice HP model with side-chains for protein folding problem

Ivan Todorin[1,*], *Nicola* Yanev [2], *Metodi* Traykov[3], *Borislav* Yurukov[4]

[1]Department of Communication and Computer Engineering, Faculty of Engineering, University Center for Advanced Bioinformatics Research, South-West University "Neofit Rilski", 66 Ivan Mihaylov Str., 2700 Blagoevgrad, Bulgaria
[2]Institute of Mathematics and Informatics, Bulgarian Academy of Sciences, Acad. Georgi Bonchev Str., Block 8, 1113 Sofia, Bulgaria
[3]Department of Electrical Engineering, Electronics and Automatics, Faculty of Engineering, University Center for Advanced Bioinformatics Research, South-West University "Neofit Rilski", 66 Ivan Mihaylov Str., 2700 Blagoevgrad, Bulgaria
[4]Department of Informatics, South-West University "Neofit Rilski", 66 Ivan Mihaylov Str., 2700 Blagoevgrad, Bulgaria

Abstract. A HP like nonlinear programming model and a "brutforce" algorithm is proposed. The model takes into account highest degree of complexity – different size of radiuses of side-chains, mentioned below as radicals, and two stages of the algorithm, including random structure initially and subsequent purposeful folding process according to the physical forces and energies. The very preliminary computational runs favourably demonstrate the adequateness of the model and the efficiency of the algorithm.

1 Introduction

3D structure of proteins is the major factor that determines their biological activity. The synthesis of new proteins and the crystallographic analysis of their 3D structure is very slow and very expensive process. If we can predict the 3D structure of many proteins, than only proteins with expected properties have to be synthesized. That will increase the number of known structures in the databases for proteins, and they can be used for drug design. The prediction of the 3D structure of proteins, if we know only the primary structure – the amino-acid sequence, is a protein folding problem. The reason for this process of folding in water environment is the interaction between water molecules and between amino-acids and water molecules. As water molecule has higher polarity than amino-acids, there is a minimum of energy when the protein is folded, not to spoil water to water interconnections. The way of folding is determined by the polarity or the hydrophobicity of different amino-acids, so the 3D structure with minimum energy is the real case. [1,2,3] There is less energy when more hydrophobic (H) amino-acids (the hydrophobic type of amino acid depends on the middle nucleotide of the codon [4]) are in contact in the core of the folded 3D structure and more polar (P) amino-acids are in contact with water. As we know the amino-acid sequence and the hydrophobicity of every amino-acid, we can predict the 3D structure – this method is called HP folding [5]. The closest to our model type of HP model is the off-lattice one, following the approach know as HP folding.

2 Methodology

2.1 Description of the set of feasible solutions

Structure and contact defining constraints

Let x_i, y_i, $z_i \in$ R be the unknown coordinates of the alpha carbon atoms, xr_i, yr_i, $zr_i \in$ R be the coordinates of the centers of the radicals and r_i be the normalized radiuses of the radicals, for the i^{th}(j^{th}) amino acid in the peptide chain, where n is the number of amino acids:

$$0.9 \le \sqrt{(x_i - x_{i+1})^2 + (y_i - y_{i+1})^2 + (z_i - z_{i+1})^2} \le 1.1$$

$$\sqrt{(x_i - xr_{i+1})^2 + (y_i - yr_{i+1})^2 + (z_i - zr_{i+1})^2} \le 1.1(r_i + 0.3)$$

$$0.6 \le \sqrt{(x_i - x_j)^2 + (y_i - y_j)^2 + (z_i - z_j)^2}$$

(1)

$$0.9(r_i + r_j) \le \sqrt{(xr_i - xr_j)^2 + (yr_i - yr_j)^2 + (zr_i - zr_j)^2}$$

$$0.9(r_j + 0.3) \le \sqrt{(x_i - xr_j)^2 + (y_i - yr_j)^2 + (z_i - zr_j)^2} \le 1.1(r\_j + 0.3)$$

The constraints below are on corresponding euclidean distances between i[th] and j[th] alpha carbon atoms and centers of radicals:

$$\sqrt{(x_i - x_j)^2 + (y_i - y_j)^2 + (z_i - z_j)^2} \le 0.8$$

(2)

$$\sqrt{(xr_i - xr_j)^2 + (yr_i - yr_j)^2 + (zr_i - zr_j)^2} \le 1.2(r_i + r_j)$$

The goal of the willpower folding is to maximize the contacts between the radicals of hydrophobic amino

* Corresponding author: itodorin@gmail.com

acids and the contacts between the alpha carbon atoms of all amino acids and the radicals with electrical charge not in contact – to find the maximum of the following objective function:

$$(3)\ F(fold) = \sum_{i\ contact\ j\ redicals} (h_i + h_j + wtw) + \sum_{i\ contact\ j\ \alpha\ carbons} wtw,$$

where h_i is the hydrophobicity value of the amino acid, wtw is a parameter for the influence of hydrogen bonds.

2.2 Model description

The main purpose of the developed algorithm is to create a structure with low potential energy, starting from randomly folded form, instead of looking for the best structure among wholly randomly generated forms. Each amino acid is represented as the position of the alpha carbon atom and the position of the center of the radical with three-dimensional coordinates. The first stage is to generate a three-dimensional shape by randomly turning the peptide chain in 90 degrees and a distance 1 between the alpha carbon atoms in the peptide chain and between each alpha carbon and the radical center of the same amino-acid. For prevention of making rather spread or extreamly tight, there is used variable constrain of spreading [6]. The second stage, which is called "Willpower folding" [7], is to purposefully modify the 3D in order to minimize the energy while using now non-integer coordinates.

2.3 Willpower folding

The steps of the algorithm for maximizing (3) subject to (1) and (2) are given below:

1. Locate the conditional center as the arithmetic mean of the three directions:
$$x_c = \sum_{i=1}^{n} \frac{xr_i}{n};\ y_c = \sum_{i=1}^{n} \frac{yr_i}{n};\ z_c = \sum_{i=1\ldots n} \frac{zr_i}{n}$$
2. Displace the coordinates of the radicals in proportion to their hydrophobicity value h_i, if the $h_i > 0$ direction is to the center of the molecule, and if $h_i < 0$ is opposite: $x_{ni} = x_i+0.02h_i$, $x_c > x_i$; $x_{i\_new} = x_i-0.02h_i$, $x_c<x_i$; $y_{i\_new} = y_i+0.02h_i$, $y_c > y_i$; $y_{i\_new} = y_i-0.02h_i$, $y_c < y_i$; $z_{i\_new} = z_i+0.02h_i$, $z_c > z_i$; $z_{i\_new} = x_i-0.02h_i$, $z_c < z_i$;
3. Displace the coordinates of the alpha carbon atoms to the center of the molecule: $x_{i\_new} = x_i+0.01$, $x_c > x_i$; $x_{i\_new} = x_i-0.01$, $x_c < x_i$; $y_{i\_new} = y_i+0.01$, $y_c > y_i$; $y_{i\_new} = y_i-0.01$, $y_c<y_i$; $z_{i\_new}=z_i+0.01$, $z_c > z_i$; $z_{i\_new} = x_i-0.01$, $z_c<z_i$;
4. Correct the positions of all alpha carbon atoms and radicals in order to preserve the peptide chain, such as alpha-carbon i approaching $i + 1$ and radical i approaching the alpha carbon atom i if $\sqrt{(x_i - x_{i+1})^2 + (y_i - y_{i+1})^2 + (z_i - z_{i+1})^2} \leq 1.1$, $x_{i\_new} = x_i + (x_{i+1}-x_i)/10$; $y_{i\_new} = y_i + (y_{i+1}-y_i)/10$; $z_{i\_new} = z_i + (z_{i+1}-z_i)/10$; $xr_{i\_new} = xr_i + (x_i-xr_i)/10$; $yr_{i\_new} = yr_i + (y_i-yr_i)/10$; $zr_{i\_new} = zr_i + (z_i-zr_i)/10$;
5. Separate the coordinates of the radicals of electrically charged amino acids from one another if $\sqrt{(xr_i - xr_j)^2 + (yr_i - yr_j)^2 + (zr_i - zr_j)^2} \leq 2$, $xr_{i\_new} = xr_i-(xr_j-xr_i)/10$; $yr_{i\_new} = yr_i-(yr_j-yr_i)/10$; $zr_{i\_new} = zr_i-(zr_j-zr_i)/10$; $xr_{j\_new} = xr_j-(xr_i-xr_j)/10$; $yr_{j\_new} = yr_j-(yr_i-yr_j)/10$; $zr_{j\_new} = zr_j-(zr_i-zr_j)/10$;
6. Reduce the distance radical coordinates of closely spaced hydrophobic amino acids if $\sqrt{(xr_i - xr_j)^2 + (yr_i - yr_j)^2 + (zr_i - zr_j)^2} \leq 2$, $xr_{i\_new} = xr_i+(xr_j-xr_i)/20$; $yr_{i\_new} = yr_i+(yr_j-yr_i)/20$; $zr_{i\_new} = zr_i+(zr_j-zr_i)/20$; $xr_{j\_new} = xr_j+(xr_i-xr_j)/20$; $yr_{j\_new} = yr_j+(yr_i-yr_j)/20$; $zr_{j\_new} = zr_j+(zr_i-zr_j)/20$;
7. Reduce the distance the coordinates of the alpha carbon atoms of closely spaced amino acids if $\sqrt{(xr_i - xr_j)^2 + (yr_i - yr_j)^2 + (zr_i - zr_j)^2} \leq 2$, $xr_{i\_new} = xr_i+(xr_j-xr_i)/20$; $yr_{i\_new} = yr_i+(yr_j-yr_i)/20$; $zr_{i\_new} = zr_i+(zr_j-zr_i)/20$; $xr_{j\_new} = xr_j+(xr_i-xr_j)/20$; $yr_{j\_new} = yr_j+(yr_i-yr_j)/20$; $zr_{j\_new} = zr_j+(zr_i-zr_j)/20$;
8. Adjust the positions of all alpha carbon atoms to avoid overlapping if $0.6 > \sqrt{(x_i - x_j)^2 + (y_i - y_j)^2 + (z_i - z_j)^2}$ as $j \neq i+1$, and $0.9 > \sqrt{(x_i - x_j)^2 + (y_i - y_j)^2 + (z_i - z_j)^2}$ as $j = i+1$, $x_{i\_new} = x_i-0.1(x_j-x_i)/((x_j-x_i)^2 + (y_j-y_i)^2 + (z_j-z_i)^2)$; $y_{i\_new} = y_i-0.1(y_j-y_i)/((x_j-x_i)^2 + (y_j-y_i)^2 + (z_j-z_i)^2)$; $z_{i\_new} = z_i-0.1(z_j-z_i)/((x_j-x_i)^2 + (y_j-y_i)^2 + (z_j-z_i)^2)$;
9. Correct the positions of all radicals to avoid overlapping if $0.9(r_i + r_j) > \sqrt{(xr_i - xr_j)^2 + (yr_i - yr_j)^2 + (zr_i - zr_j)^2}$, $xr_{i\_new} = xr_i-0.1(xr_j-xr_i)/((xr_j-xr_i)^2 + (yr_j-yr_i)^2 + (zr_j-zr_i)^2)$; $yr_{i\_new} = yr_i-0.1(yr_j-yr_i)/((xr_j-xr_i)^2 + (yr_j-yr_i)^2 + (zr_j-zr_i)^2)$; $zr_{i\_new} = zr_i-0.1(zr_j-zr_i)/((xr_j-xr_i)^2 + (yr_j-yr_i)^2 + (zr_j-zr_i)^2)$;
10. Adjust positions to avoid overlapping between the alpha carbon atom and the radical, if $0.9(r_j + 0.3) > \sqrt{(xr_i - xr_j)^2 + (yr_i - yr_j)^2 + (zr_i - zr_j)^2}$, $xr_{i\_new} = xr_i-0.1(xr_j-x_i)/((xr_j-r_i)^2 + (yr_j-y_i)^2 + (zr_j-z_i)^2)$; $yr_{i\_new} = yr_i-0.1(yr_j-y_i)/((xr_j-x_i)^2 + (yr_j-y_i)^2 + (zr_j-z_i)^2)$; $zr_{i\_new} = zr_i-0.1(zr_j-z_i)/((xr_j-r_i)^2 + (yr_j-r_i)^2 + (zr_j-z_i)^2)$;

2.4 Thermo effect

What we call "Thermo effect" is to make a small random move in every step of Willpower folding, which corresponds to the real environment. This prevents building equal structures if the initial random fold is the same and gives more chance to find better structure.

3 Computational Results

First, using our definition of contact (2), we find the contacts in the real structure of protein 1UUB, using the coordinates of alpha carbons in PDB file of the protein data bank (Figure 1).

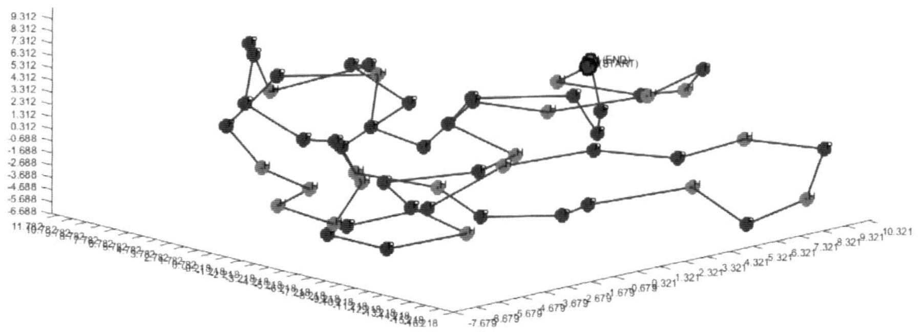

Fig. 1. Real structure of protein 1UUB

There are 125 contacts according to the constraints (2) above.

The test run of the program, realizing the algorithm on 1UUB obtains the following results: among only 100000 randomly generated structures, the value of the evaluation function is 1279.2 and 206 contacts – 59 matches with the real one (Figure 2).. The time needed was 70 min on PC, Intel i5, 2.26GHz.

Remark: for the source code in C++, the list of values coordinates and the list of contacts' couples, mail to itodorin@gmail.com.

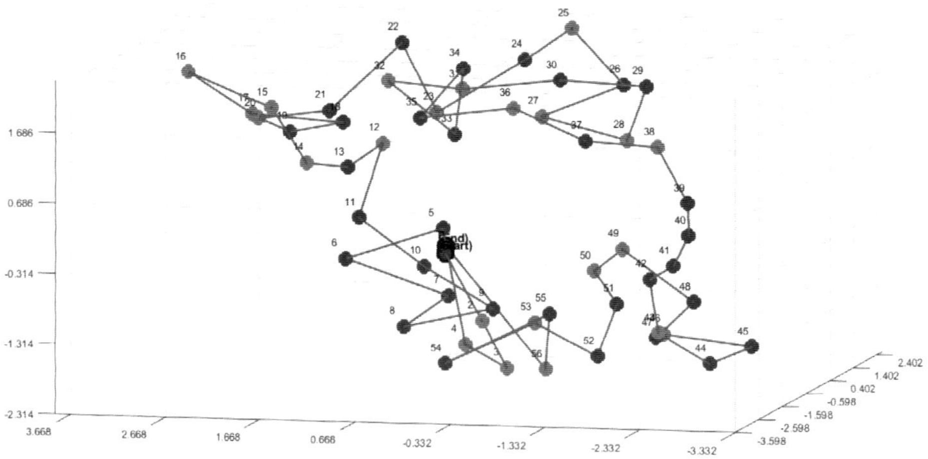

Fig. 2. The structure with 59 matches

The other experiment we done, is to find out how much better structure could be built, classifying fold not by evaluation function, but by ratio contacts matching ratio. In this way we obtain the following structure with 62 contacts matches out of 120 (Figure 3).

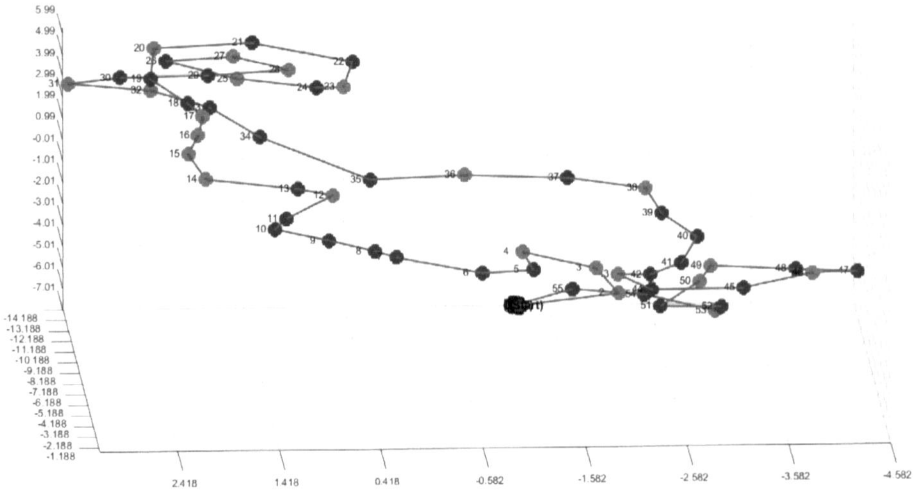

Fig. 3. The structure with 62 contacts matches

4 Conclusions

This approach for protein folding prediction, using "Willpower folding" with "Thermo effect", appear to be faster than models for finding best structure among randomly built structures. Therefore it is possible rather more structures of proteins to be processed for the same time. Even more, the model is new (age under year) and might be further developed in direction of granularity – placing every atom, and the evaluation function could take into account the real impact of the distribution of electronic density, such degree of granularity is hard to be execute considering computational time for models for finding best structure among randomly built structures.

5. Acknowledgements

This work is partially supported by the project of the Bulgarian National Science Fund, entitled: "Bioinformatics research: protein folding, docking and prediction of biological activity", code NSF I02/16, 12.12.14.

References

1. A. Kolinski and J. Skolnick, Proteins **18**, 338-352 (1994).
2. B. Berger, and T. Leighton, Journal of Computational Biology **5**, 27-40 (1998).
3. H. Greenberg, W. Hart, G. Lancia, INFORMS Journal on Computing **16**, 211-231 (2004).
4. P. Milanov, I. Trenchev, N. Pencheva, Mathematica Balkanica **17**, 157-165 (2004).
5. S. Istrail, and F. Lam, Commun. Inf. Syst. **9**, 303-346 (2009).
6. I. Todorin, Proceedings of the Fifth International Scientific Conference – FMNS2013 **1**, 188-192 (2013).
7. I. Todorin, SDNS **1**, 1-6 (2017) (in printing).

Computer modeling of Cannabinoid receptor type 1

Fatima Sapundzhi[1,*], *Tatyana* Dzimbova [1,2], *Nevena* Pencheva[1] and *Peter* Milanov[1,3]

[1]South-West University "Neofit Rilski", Bulgaria, 2700 Blagoevgrad
[2]Institute of Molecular Biology, Bulgarian Academy of Sciences, Bulgaria, 1113 Sofia
[3]Institute of Mathematics and Informatics, Bulgarian Academy of Sciences, Bulgaria, 1113 Sofia

Abstract. Cannabinoid receptors are important class of receptors as they are involved in various physiological processes such as appetite, pain-sensation, mood, and memory. It is important to design receptor-selective ligands in order to treat a particular disorder. The aim of the present study is to model the structure of cannabinoid receptor CB1 and to perform docking between obtained models and known ligands. Two models of CBR1 were prepared with two different methods (Modeller of Chimera and MOE). They were used for docking with GOLD 5.2. It was established a high correlation between inhibitory constant Ki of CB1 cannabinoid ligands and the ChemScore scoring function of GOLD, which concerns both models. This suggests that the models of the CB1 receptors obtained could be used for docking studies and in further investigation and design of new potential, selective and active cannabinoids with the desired effects.

1 Introduction

Cannabinoid receptors (CBR) are class of transmembrane receptors part of G protein-coupled receptors (GPCRs). They are located mainly in the brain and are involved in various physiological processes such as appetite, pain-sensation, mood, and memory [1, 2].

CBR can be activated by three major groups of ligands, endocannabinoids (produced by the mammalian body), phytocannabinoids (produced by the cannabis plant) and synthetic cannabinoids.

Two subtypes of CBR are currently known, cannabinoid receptor type 1 (CB1) and cannabinoid receptor type 2 (CB2) [3, 4]. The CB1 receptor is expressed mainly in the brain, but also in the lungs, liver and kidneys. The CB2 receptor is expressed in the immune system and in hematopoietic cells. Begg et al. evidence suggests that there are novel cannabinoid receptors [5], which are expressed in endothelial cells and in the CNS.

The protein sequences of CB1 and CB2 receptors are about 44% similar [6, 7]. Cannabinoids bind reversibly and stereo-selectively to the CBR. The CB1 receptors are thought to be one of the most widely expressed GPCRs in the brain. They are also found in other parts of the body. For instance, in the liver, activation of the CB1 receptor is known to increase *de novo* lipogenesis [8].

It is very important to develop subtype selective cannabinoid ligands in order to treat different diseases, including obesity. Docking studies carried out with various groups of cannabinoids and with cannabinoid receptors, would allow to establish the structure-activity relationships, to search for potent analogues and to understand drug-receptor interactions. But the crystal structures of the CBR are still not available in the data bases.

The aim of the present study is to build a model of CB1 receptor by homology modelling in order to use it in further research.

2 Methods

The amino acid sequence of the cannabinoid receptor was obtained from UniProtKB (P21554) [9]. 3D structure of the CB1 receptor was generated by using two different methods.

The first method consists of several steps that are held on different software. The structure of rhodopsin (PDB id: 2Z73, crystal structure of squid rhodopsin) was used as template [10]. The alignment of the two sequences was performed in CLUSTALW and the obtained alignment file was used in Chimera for generating 3D-structure of the receptor. Five structures were generated and the structure with the best GA341 was chosen and additionally analyzed with PROCHECK [11] and MolProbity [12].

The second method uses the software MOE (Molecular Operating Environment) in direct generation of the receptor structure [13]. The software MOE is a

* Corresponding author: sapundzhi@swu.bg

chemical computing and molecular modelling tool, which is a very widely used program in scientific applications. The steps included: searching for the template in the data base (the structure with the highest per cent of identity of the amino acid residues), modeling of the 3D structure of the desired receptor and checking the obtained model.

GOLD (Genetic Optimisation for Ligand Docking) 5.2 [14] was used for the molecular docking studies of the obtained two models of CB1 receptors by homology modelling with the ligands known from the literature, presented in Table 1 [15-18].

Table 1. The values of the inhibitory constant (Ki) of CB1 cannabinoid receptor ligands with known affinity, used in the docking studies.

Ligands	**CB1 Affinity (K_i)**
Anandamide	78 nM
N-Arachidonoyl dopamine	-
2-Arachidonoylglycerol	-
2-Arachidonyl glyceryl ether	21 nM
Δ-9-Tetrahydrocannabinol	10 nM
EGCG (Epigallocatechin Gallate)	33.6 μM
Yangonin	0.72 μM
UR-144	150 nM

GOLD 5.2 uses generic algorithm and considers full ligand conformational flexibility and partial protein flexibility. All scoring functions available in the tool are used: GoldScore, ChemScore, ChemPLP (Piecewise Linear Potential), and ASP (Astex Statistical Potential) scoring functions, [19-23]. GoldScore fitness function is a molecular mechanics–like function (Eq.1):

$$GoldScore = S_{hb\_ext} + S_{vdw\_ext} + S_{hb\_int} + S_{vdw\_int} \quad (1)$$

where S_{hb_ext} is protein-ligand hydrogen bond energy, S_{vdw_ext} - protein-ligand van der Waals energy, S_{hb_int} is intramolecular hydrogen bonds in the ligand and S_{vdw_int} is intramolecular strain in the ligand [21].

$$ChemScore = \Delta G_{binding} + E_{clas\,h} + E_{int} + E_{cov} \quad (2)$$

ChemScore scoring function takes count of hydrophobic-hydrophobic contact area, ligand flexibility, hydrogen bonding and metal interaction (Eq.2) [21], where $\Delta G_{binding}$ is the free energy of binding of the ligand to the protein; $E_{clas\,h}$ is a protein-ligand clash-energy term, E_{int} - ligand-internal-energy term.

ChemPLP is the most effective scoring function for pose prediction in cognate protein-ligand complexes, out of the four available in GOLD [20] (Eq. 3):

$$f_{CHEMPLP} = f_{PLP} - (f_{chem-hb} + f_{chem-cho} + f_{chem-met}) \quad (3)$$

where f_{PLP} is the modelling of the steric complementarity between protein and ligand, $f_{chem-hb}$ is a distance-dependent hydrogen bonding term, $f_{chem-cho}$ is an angle-dependent hydrogen bonding term, $f_{chem-met}$ is a distance-dependent metal bonding term.

ASP scoring function is an atom-atom distance potential derived from a database of protein-ligand complexes and can be compared to other knowledge-based scoring potentials [22] (Eq. 4):

$$ASP\ Fitness = -C_s S(map) - c_{int} E_{int} - c_{clas\,h} E_{clas\,h} \quad (4)$$

where C_s is a scaling factor, $S(map)$ - total statescore - all combinations of protein and ligand atoms, c_{int} - internal energy, $c_{clas\,h}$-clash coefficients, $E_{clas\,h}$ and E_{int} are terms like in ChemPLP function.

In order to find relationship between the affinity of cannabinoid ligands (Table 1) and the docking results for the obtained models of CB1 receptors by homology modelling, we tried to predict it with the help of Pearson's correlation, using GraphPad Prism 3.0 (http://www.graphpad.com/scientific-software/prism).

3 Results and discussion

By using the first method for generation of the 3D model of CB1 receptor, five models were obtained. The model with the best value of GA345 was chosen for further analysis.

Evaluations of 3D models were done for various levels of structural organization and on the base of *in vitro* studies. We generate 40 models using Modeller of Chimera. One of them with best GA345 (more than 0.9) was chosen for further evaluation.

The geometry of the obtained model, analyzed with PROCHECK and MolProbity, and the respective Ramachandran plots are presented in Figure 1.

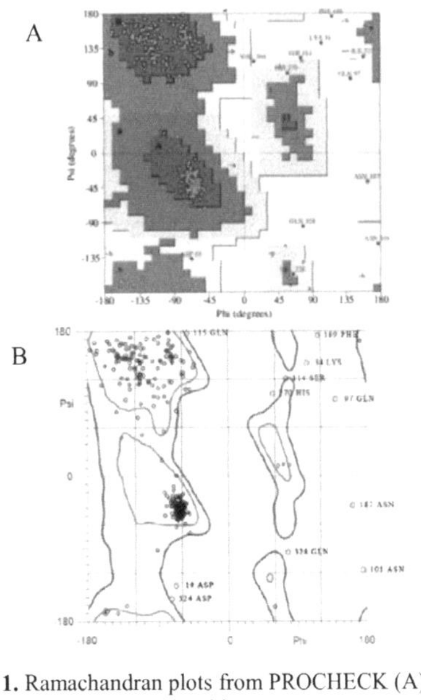

Fig. 1. Ramachandran plots from PROCHECK (A) and MolProbity (B).

The results are good as the sum of the most favored and additional allowed regions is more than 90% - 97.3% in PROCHECK analysis [11] and 96.6% in MolProbity analysis [12].

https://doi.org/10.1051/itmconf/20181602008

The second model was obtained with the help of MOE software [13]. The program automatically found the structure similarity with the same template used in the first method (the crystal structure of squid rhodopsin (PDBid:2z73) [10]). The standard molecular mechanics AMBER 99 forcefield was used. The backbone fragments from a high-resolution structural database were collected and alternative side chain conformations for non-identical residues were assembled from an extensive rotamer library during data collection. During model building in software MOE [13], ten independent models were created based upon loop and side chain placements scored by a contact energy function (Figure 2).

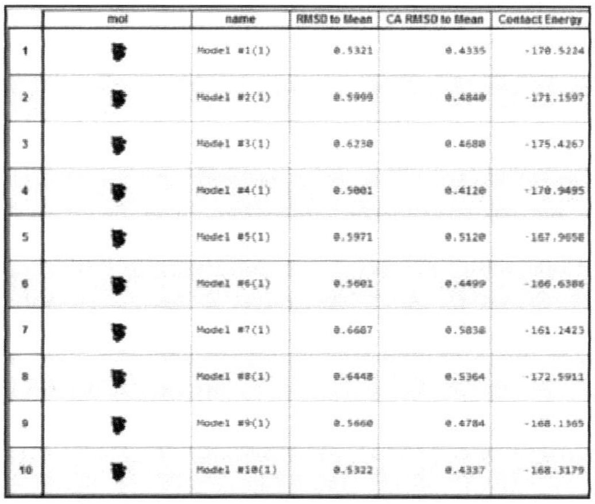

	mol	name	RMSD to Mean	CA RMSD to Mean	Contact Energy
1		Model #1(1)	0.5321	0.4335	-170.5224
2		Model #2(1)	0.5909	0.4840	-171.1597
3		Model #3(1)	0.6230	0.4680	-175.4267
4		Model #4(1)	0.5001	0.4120	-170.9495
5		Model #5(1)	0.5971	0.5120	-167.9658
6		Model #6(1)	0.5601	0.4499	-166.6386
7		Model #7(1)	0.6687	0.5838	-161.2423
8		Model #8(1)	0.6448	0.5364	-172.5911
9		Model #9(1)	0.5660	0.4784	-168.1365
10		Model #10(1)	0.5322	0.4337	-168.3179

Fig. 2. Screenshot of database obtained after modelling by MOE.

The final, refined model was loaded in MOE and Chimera after the calculation finished. Graphical representations of the obtained model of homology modelling of CB1 are presented in Figure 3.

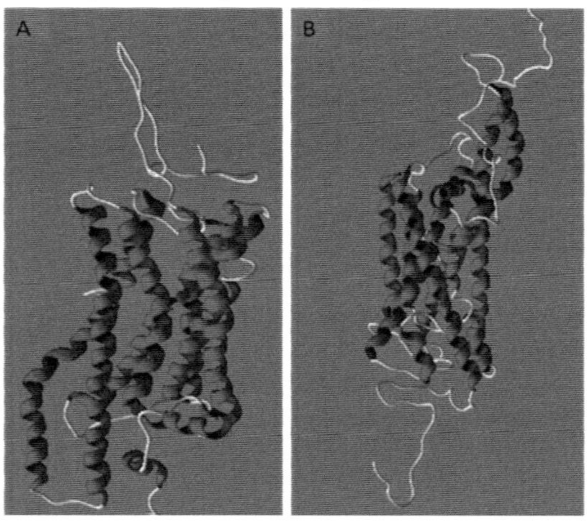

Fig. 3. Graphical representation of the obtained model of homology modelling of CB1 by the software MOE (A) and the software Chimera (B).

After that, the stereochemistry of homology models for unusual or geometrically unreasonable features was examined. Our model has RMSD 0.6230 < 3A°, which means that the generated model is built correctly.

The model with the best contact energy was chosen (-175,4267) and it was validated directly in MOE. The stereochemical quality of the modeled proteins is assessed from Ramachandran validation score for favoured regions and allowed regions (Figure 4). In general, a score close to 100% implies good stereochemical quality of the models.

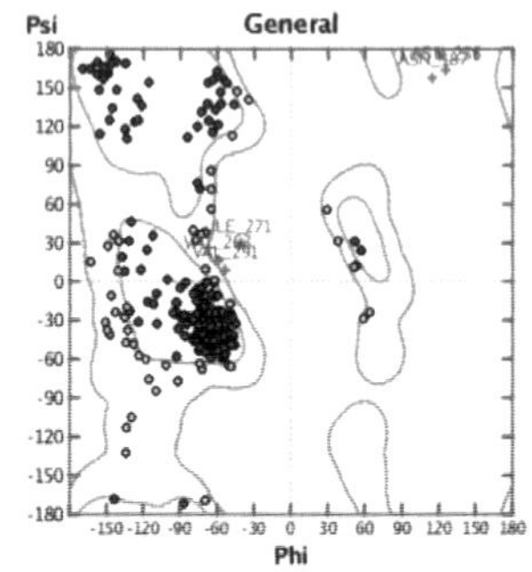

Fig. 4. Ramachandran Plot generated by MOE after optimization of the structure.

The molecular docking experiments with the obtained models of CB1 receptor by homology modelling in Chimera and in MOE and the ligands from the literature (Table 1) were carried out with software GOLD 5.2 and all four scoring functions embedded in the program: GoldScore, ChemScore, ASP and ChemPLP [19-23]. According to Shim et al. [24] there exists a hydrophobic binding pocket that interacts with the alkyl chain of the cannabinoids. The docking is effective when the polar residue from the receptor sequence was chosen - Asp366 and the investigated ligands bind near to it.

Correlations between the values of the respective GOLD functions, obtained for the models of CB1 receptor with Modeller of Chimera and MOE, and the affinity constant - K_i of the cannabinoid ligands, calculated with Pearson's coefficient, are presented in Table 2. These data show that the best correlation concerns the model of CB1 receptor obtained by MOE and the scoring function ChemScore, because the value of the Pearson's coefficient is highest and the relationship is negative. For the CB1 model, generated by Chimera and the same scoring function, the correlation coefficient was similar [25-28].

Table 2. The values of Pearson's coefficient for the correlation between GOLD scoring functions for models of the CB1 receptor (Chimera and MOE; homology modeling) and affinity constant K_i of the cannabinoid ligands.

GOLD Scoring Function	Pearson correlation coefficient - R	
	CB1 model with Chimera	CB1 model with MOE
ASP	0.9248	0.9594
ChemPLP	- 0.4851	- 0.0817

https://doi.org/10.1051/itmconf/20181602008

ChemScore	- 0.9491	- 0.9944
GoldScore	- 0.3001	- 0.0651

Thus, both models of the CB1 receptors, created by Chimera or MOE are suitable for docking studies, but using only ChemScore scoring function.

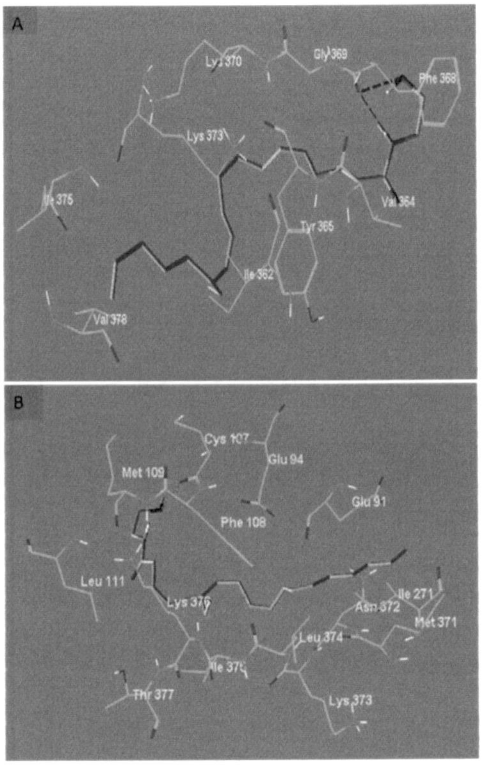

Fig. 5. Graphical representation of the docking between the cannabinoid ligand Anandamid and the ASP scoring function (A) and ChemScore scoring function (B) .

In our previous studies it was generated a 3D model of Delta opioid receptor (DOR) by using homology modelling [27]. Newly generated model of DOR was used for *in silico* studies and it was compared with recently published data for crystal structure of DOR [28-31]. It was established that the most appropriate combination for analysis of investigated compounds and DOR is that between the model of DOR obtained by homology modelling and the ASP scoring function (available in GOLD 5.2) [31-34]. These "preferences" of the opioids to this scoring function and of the cannabinoids to the ChemScore function, found in this study, need further investigations.

In statistics the negative correlation is a relationship between two variables in which one variable increases as the other decreases, and vice versa. A negative correlation between the values for obtained models by homology modelling of the CB1 receptor (Chimera and MOE) for ChemScore scoring function and the values of affinity of the cannabinoid ligands was established. In our case, these obtained negative values of Pearson's correlation coefficient have a biological meaning. As lower the value of the constant K_i is the biological effect is the stronger.

The presented work opens wide space for the design of novel cannabinod analogues, selective to the CB1 receptor and their assessment with the molecular docking studies. The obtained models of CB1 cannabinoid receptors allow the studying of the structure-activity relationship of synthetic, endogenous or plant–derived ligands that can act as agonists or antagonists to the CB1 receptor.

Conclusions

In the present study there are generated two models of the CB1 receptor, obtained with the Modeller of Chimera and MOE. The models are suitable for docking studies, because it was established a high correlation between Ki of CB1 cannabinoid ligands and the ChemScore scoring function of GOLD, which concerns both models.

Acknowledgements

This work is partially supported by the project of the Bulgarian National Science Fund, entitled: Bioinformatics research: protein folding, docking and prediction of biological activity, NSF I02/16, 12.12.14.

References

1. A.Howlett. Prostaglandins & Other Lipid Mediators **68-69**, 619–631 (2002)
2. K. Mackie. Journal of Neuroendocrinology, **20** (1), 10-14 (2008)
3. L. Matsuda, et al. Nature **346**, 561 - 564 (1990)
4. C. Gérard, C. Mollereau, G. Vassart & Parmentier. Biochem J **279**, 129–134 (1991)
5. M. Begg, et al. Pharmacology & Therapeutics **106**, 133 – 145 (2005)
6. D. Latek, et al. J Mol Model **17**, 2353-2366 (2011)
7. S. Munro, K. Thomas, M.Abu-Shaar, Nature **365**, 61-65 (1993)
8. D. Osei-Hyiaman, et al. J Clin Invest. **115**, 1298-1305 (2005)
9. http://www.uniprot.org/uniprot/P21554
10. M. Murakami, T. Kouyama, Nature **453**, 363-367 (2008)
11. R. Laskowski, M. MacArthur, D. Moss, J. Thornton, J Appl Cryst **26**, 283-291 (1993)
12. V. Chen, W. Arendall, J. Headd, D. Keedy, R. Immormino, G. Kapral, L. Murray, J. Richardson, D. Richardson, Acta Crystallogr D Biol Crystallogr, **66**, 12-21 (2010)
13. http://www.chemcomp.com/
14. G. Jones, P. Willett, R. Glen, A. Leach, R. Taylor, J Mol Biol. **267**, 727–748 (1997)
15. A. Ligresti, R. Villano, M. Allarà, I. Ujváry, V. Di Marzo, Pharmacological Research, **66** (2), 163-169 (2012)

16. A. Makriyannis, H. Deng, "Cannabimimetic indole derivatives", granted 2001-06-07
17. J. Frost et al., J. Med. Chem. **53 (1)**, 295–315 (2010)
18. M. Aung, et al., Drug Alcohol Depend. **60 (2)**, 133–140 (2000)
19. GOLD, version 5.2 UserGuide, CCDC Software Ltd.Cambridge, UK (2010).
20. C. Baxter, C. Murray W, D. Clark, D. Westhead, M. Eldridge. Proteins, **33**, 367–382 (1998)
21. O. Korb, T. Stützle, T. Exner, J. Chem, Inf. Mod., **49**, 84-96, (2009)
22. M. Verdonk, J. Cole, M. Hartshorn, C. Murray, R. Taylor, Proteins, **52,** 609-623, (2003)
23. R. Reinscheid, H. Nothacker, A. Bourson, A. Ardati, R. Henningsen, J. Bunzow, D. Grandy, H. Langen, F. Monsma, O. Civelli, Science, **270,** 792-794 (1995)
24. Joong-Youn Shim, A. Bertalovitz, D. Kendall. Journal of biological chemistry, **286 (38),** 33422–33435 (2011)
25. F. Sapundzhi, T. Dzimbova, N. Pencheva, P. Milanov, Proceedings of Biomath Communications Supplement, 25-30 June 2017, Skukuza Camp, South Africa, **4(1)**, (2017)
26. F. Sapundzhi, T. Dzimbova, N. Pencheva, P. Milanov, Proceedings of Seventh International Conference of FMNS 2017, 14-18 June 2017, Blagoevgrad, Bulgaria, **1**, (2017)
27. T. Dzimbova, F. Sapundzhi, N. Pencheva, P. Milanov, Journal of Peptide Science, **18** (S1) S84, P072, (2012)
28. F. Sapundzhi, T. Dzimbova, N. Pencheva, P. Milanov, Journal of Peptide Science, **20** (S1), S294, (2014)
29. T. Dzimbova, F. Sapundzhi, N. Pencheva, P. Milanov, Int. J. Bioautomation, **17**, 5-16 (2013)
30. F. Sapundzhi, T. Dzimbova, N. Pencheva, P. Milanov, Jurnal of Bulgarian Chemical Communication, **2**, 613-618 (2015).
31. F. Sapundzhi, T. Dzimbova, N. Pencheva, P. Milanov, Der Pharma Chemica, **8**, 118-124 (2016)
32. R. Mavrevski, C. R. Acad. Bulg. Sci., **67(10)**, 1345-1354 (2014)
33. R. Mavrevski, P. Milanov, B. Yurukov, Proceedings of the Biomath Communications – BIOMATH2016, **3(1)**, 46 (2016)
34. T. Dzimbova, R. Mavrevski, N. Pencheva, T. Pajpanova, P. Milanov, Bulgarian chemical communications, **44(3)**, 242-246 (2012)

Analysis of docking algorithms by HPC methods generated in bioinformatics studies

Anton Stoilov[1,2*], *Borislav* Yurukov [3], *Peter* Milanov [3,4]

[1] South-West University, Faculty of Engineering, Department of Electrotechnics, Electronics and Automation, Bulgaria

[2]American University in Bulgaria, Department of Computer Science, Bulgaria

[3]South-West University, Faculty of Mathematics and Natural Science, Department of Informatics, Bulgaria

[4]Bulgarian Academy of Sciences, Institute of Mathematics and Informatics, Bulgaria

Abstract. High-performance computing (HPC) is an important domain of the computer science field. For more than 30 years, it has allowed finding solutions to problems and enhanced progress in many scientific areas such as bioinformatics and drug design. The binding of small molecule ligands to large protein targets is central to numerous biological processes. The accurate prediction of the binding modes between the ligand and protein (the docking problem) is of fundamental importance in modern structure-based drug design. The interactions between the receptor and ligand are quantum mechanical in nature, but due to the complexity of biological systems, quantum theory cannot be applied directly. Consequently, most methods used in docking and computational drug discovery are more empirical in nature and usually lack generality.

1 Introduction

Quantum mechanical phenomena, such as the formation of a covalent bond between the protein and the ligand upon binding during the transition state of the reaction, cannot be predicted and/or evaluated using these empirical methods. In the field of molecular modeling, docking is a method which predicts the preferred orientation of one molecule to a second when bound to each other to form a stable complex. Knowledge of the preferred orientation in turn may be used to predict the strength of association or binding affinity between two molecules using, for example, scoring functions. Docking is frequently used to predict the binding orientation of small molecule drug candidates to their protein targets in order to in turn predict the affinity and activity of the small molecule. Hence docking plays an important role in the rational design of drugs. Given the biological and pharmaceutical significance of molecular docking, considerable efforts have been directed towards improving the methods used to predict docking. Each docking program makes use of one or more specific search algorithms, which are the methods used to predict the possible conformations of a binary complex. The present benchmark is made from an existing test set (CCDC/Astex Validation Set) on typical HPC system. Selected examples were docked with GOLD software.

2 Molecular docking

Molecular docking is a computer simulation procedure to predict the conformation of a receptor-ligand complex, where the receptor is usually a protein or a nucleic acid molecule (DNA or RNA) and the ligand is either a small molecule or another protein. It can also be defined as a simulation process where a ligand position is estimated in a predicted or pre-defined binding site. Molecular docking research focusses on computationally simulating the molecular recognition process. It aims to achieve an optimized conformation for both the protein and ligand and relative orientation between protein and ligand such that the free energy of the overall system is minimized. Computational docking of a small molecule to a biological target involves efficient sampling of possible poses of the former in the specified binding pocket of the latter in order to identify the optimal binding geometry, as measured by a user-defined fitness or score function. X-ray crystallography and NMR spectroscopy continue to be the primary source of 3-dimensional structural data for protein and nucleic acid targets. In favourable cases where proteins of unknown structure have high sequence homology to known structures, homology modelling can provide a viable alternative by generating a suitable starting point for "in silico" discovery of high affinity ligands. Potential energy of molecular field model is a function of a atomic position (x,y,z) normally in Cartesian space. The equation of the potential energy of the system of atoms in the molecular force field, commonly used in molecular modelling is presented below:

* Corresponding author: antonstoilov@swu.bg

$$E = \frac{1}{2}\sum_{i=1}^{m} k_{\theta i}(\theta_i - \theta_{0,i})^2 + \frac{1}{2}\sum_{i=1}^{n} k_{bi}(b_i - b_{0,i})^2 + \frac{1}{2}\sum_{i=1}^{k} v_i\left[1+\cos(n_i\omega_i - \gamma_i)\right] + \sum_{i,j} 4\varepsilon_{i,j}\left[\left(\frac{\sigma_i}{r_{ij}}\right)^{12} - \left(\frac{\sigma_j}{r_{ij}}\right)^{6}\right] + \sum_{i,j}\frac{1}{4\pi\varepsilon_0}\frac{q_i q_j}{r_{ij}^2} \quad (1)$$

The complexity of computational docking increases in the following order:
(a) rigid body docking, where both the receptor and small molecule are treated as rigid. (b) flexible ligand docking, where the receptor is held rigid, but the ligand is treated as flexible; (c) flexible docking, where both receptor and ligand flexibility is considered.
Rigid-body docking simulation has been employed for virtual-screening initiatives, this method has been used as the fastest way to perform an initial screening of a small molecule database. It has a relatively high accuracy, when compared against crystallographic structures. This accuracy is even higher if we introduced an analysis of the best results using an empirical scoring function for the best results obtained using rigid-body docking simulations. Usually, flexible docking or/and scoring functions have been used for applying a more specific refinement and lead optimization after initial rigid body docking procedure, since these methods demand for computational power and CPU time. Flexible docking methods can consider several possible conformations of ligand or receptor, as well as for both molecules at the same time, at a higher computational time cost. The root-mean-square deviation (RMSD) is calculated between two sets of atomic coordinates, in this case, one for the crystallographic structure (xc, yc, zc) and another for the atomic coordinates obtained from the docking simulations (xd, yd, zd), the summation is taken over all N atoms being compared, the equation is as follows:

$$RMSD = \sqrt{\frac{1}{N}\sum_{i=1}^{N}(x_{ci} - x_{di})^2 + (y_{ci} - y_{di})^2 + (z_{ci} - z_{di})^2} \quad (2)$$

Before ligands can be docked against a receptor, generally the binding site has to be identified first. This is done to limit the search space on the receptor surface and thus minimize the degrees of freedom that have to be searched. The active site is often known from crystal structures of ligand-bound receptors, but it can also be predicted. The largest cavity n a protein surface is frequently the active site, but this is not always the case and different active site prediction and analysis methods have been developed.

The genetic algorithm (GA) adopted by GOLD algorithm requires as input the approximate size and location of the receptor active site and also the coordinates of protein and a ligand conformation. The active site may be defined by several techniques. GA is also implemented in the program DOCK, which is able to dock either whole ligand inside active site or a rigid fragment of the ligand."Lamarckian" GA (LGA) is also implemented in docking algorithms. The LGA switches between "genotypic space" and "phenotypic space." Mutation and crossover occur in genotypic space, while phenotypic space is determined by the energy function to be optimized. Energy minimization (local sampling) is performed after genotypic changes have been made to the population (global sampling) in phenotypic space, which is conceptually similar to MC minimization. After successful docking procedure most important parameters are:

- Binding Energy - (BE), kcal/mol
- Intermol Energy – (ImolE), kcal/mol
- Internal Energy – (IE), kcal/mol
- Torsional Energy – (TE), kcal/mol
- Unbound Energy (UE), kcal/mol

Binding energy calculations can be performed either concurrently with ligand docking or separately for a predetermined protein-ligand structure, obtained from experimental data or other molecular modeling studies. The Binding energy is calculated by sum from Intermol Energy, Internal Energy, Torsional Energy and Unbound Energy.

(3) BE = ImolE + IE + TE + UE

3 Methods

GOLD software uses a Genetic Algorithm (GA) for protein ligand docking which works as follows:

1. Selecting a Protein
2. Adding Hydrogen Atoms
3. Deleting Waters
4. Defining the Protein Binding Site
5. Specifying the Ligand
6. Selecting a Fitness Function
7. Starting the Docking Run
8. Analysis of Output

The population of chromosomes is iteratively optimised. At each step, a point mutation may occur in a chromosome, or two chromosomes may mate to give a child. The selection of parent chromosomes is biased towards fitter members of the population, i.e. chromosomes corresponding to ligand dockings with good fitness scores. The GOLD validation test set is one of the most comprehensive of all of the docking methods reviewed, and achieved a 71% success rate based primarily on a visual inspection of the docked structures. 66 of the complexes had an RMSD of 2.0 Å or less, while 71 had an RMSD of 3.0 Å or less. The omission of hydrophobic interactions and a solvent model may explain some of the docking failures which included highly flexible, hydrophobic ligands, and those complexes containing poorly resolved active sites. However, recent extensions to GOLD include the addition of hydrophobic fitting points that are used in the least squares fitting algorithm to generate the ligand orientation.

In this paper the benchmark test set is based on CCDC/Astex Validation Set developed by Cambridge Crystallographic Data Centre (CCDC) for docking software GOLD. There are 60 entries and protonation states have been set in all cases.

The equipment for experiments was provided by CENTER FOR ADVANCED BIOINFORMATICS RESEARCH, South-West University "Neofit Rilski", Blagoevgrad, BULGARIA. This equipment include two different computational power with this characteristics:
IBM x3650 M2 – Processor: 2 x Xeon Quad-Core Intel Xeon 4C Model E5520 4C 2.26GHz with EM64T/1066MHz, 8MB L3 Cache, RAM: 2 x2GB 4096MB ECC PC3-10600 DDR3, HDD: 2x146GB 10K SAS Hot Swap RAID controller - Integrated RAID MR10i

IBM BladeCenter HS22 – Processor: 2 x Xeon Quad-Core Intel Xeon 4C Model E5504 80W 2.00GHz/800MHz/4MB L2, 2x2GB, O/Bay 2.5in SAS 1, RAM: 4 x 2GB Single Rank PC3-10600 CL9 ECC DDR3, HDD: 2 x IBM 146 GB 2.5in SFF Slim-HS 10K 6Gbps SAS HDD

4 Results

After the execution of 60 docking procedure were obtained 10 conformation for each case. After that from those 10 confirmation is chosen most energy-favorable i.e. that have the smallest binding energy (BE) so this is the main criteria for present results in Table 1. All confirmation are saved and observed. The summary of these most energy-favorable conformation for each test case are presented in Table 1.

Table 1. The results from docking procedure for each test case.

Tab. 1. The results from docking procedure for each test-case

ID of protein complex	Binding Energy, kcal/mol	Unbound Energy, kcal/mol	ID of protein complex	Binding Energy, kcal/mol	Unbound Energy, kcal/mol	ID of protein complex	Binding Energy, kcal/mol	Unbound Energy, kcal/mol
1a07	-2.91	-1.92	**1b6n**	-7.99	-1.27	**1d3h**	202.73	-0.49
1a0q	-5.17	-0.31	**1b9v**	-1.13	-1.04	**1dbm**	-7.87	-0.5
1a1b	-0.66	1.59	**1bbp**	-11.03	0.76	**1dd7**	55.69	50.78
1a1e	-6.04	-0.58	**1bgo**	253.61	0.14	**1dg5**	-6.44	-0.19
1a28	-9.26	-0.07	**1bl7**	-5.17	0.23	**1dhf**	-5.45	1.77
1a4g	0.87	0.19	**1bmq**	830.62	32.7	**1dmp**	-11.19	-2.39
1a4k	-6.36	-0.8	**1c12**	-5.78	-0.47	**1dog**	-6.75	-0.19
1a4q	16.59	-1.56	**1c1e**	-8.07	0	**1dwb**	-5.35	0.12
1a6w	-2.72	4.51	**1c2t**	18.01	2.56	**1ebg**	-8.35	0.57
1a9u	-5.78	-0.51	**1c5c**	-5.77	0.44	**1eed**	4.29	6.39
1acj	-7.81	0.1	**1c5x**	-1.33	0.1	**1eil**	-1.48	1.95
1aco	-9.03	1.25	**1c83**	62.59	-0.11	**1ejn**	113.67	1.72
1aha	-4.3	0.07	**1ckp**	-4.22	0.03	**1elb**	151.07	83.18
1ai5	-4.79	0.13	**1cps**	9.89	9.87	**1elc**	737.82	38.41
1aj7	-3.12	0.23	**1cqp**	-5.32	-1.4	**1eld**	171.57	57.6
1ake	5.67	6.59	**1ctr**	-6.48	-0.41	**1ele**	11.03	1.07
1apt	-6.02	1.16	**1ctt**	0.27	0.2	**1eoc**	-2.25	-0.18
1apu	-2.92	6.06	**1cvu**	-7.04	-0.92	**1epo**	13.27	19.43
1azm	-4.76	-0.15	**1cx2**	-10.93	0.21	**1ett**	284.34	3.49
1b58	-8.69	0.27	**1d0l**	-5.95	-1.71	**1etz**	-6.85	-2.57

The ligand will have been docked a number of times so a set of files will have been written to the output directory, each containing the results of a separate docking attempt. The result of each docking attempt is written out as gold_soln_ligand_m1_n.mol2, where n is the number of the docking solution 1,2,3 ... and m1 is an index to the ligand (in this example, only one ligand was docked). Note that the file gold_soln_ligand_m1_1.mol2 is not the best GOLD prediction, it is just the solution found in the first docking attempt. However, as GOLD proceeds, symbolic links are created: ranked_ligand_m1_1.mol2 will point to the current top-ranked solution, ranked_ligand_m1_2.mol2 will point to the second-best solution, and so on. With the Hermes 3D view is possible to inspect the solutions predicted by GOLD. The docking solutions are given in their docked order with their corresponding fitness score. If required the solutions can be ordered and Fitness to determine which is the highest scoring. A simple test of the effectiveness of a docking program is to take a protein-ligand complex from the PDB and extract the ligand. The docking program can then be used to predict the binding mode of the ligand and a comparison made with the crystallographically observed position. The crystallographically observed conformation of the docked ligand is stored in the ligand we extracted from the protein, that was subsequently re-loaded. Compare this with the solution predicted by GOLD.

After successful docking procedure for each test-case is observed by Fitness (scoring function), Best ranking time and Total run time . The results for 22 faster cases for docking from initial 60 experiments was present on Table 2. In the fields of molecular modelling, scoring functions are fast approximate mathematical methods used to predict the strength of the non-covalent interaction (also referred to as binding affinity) between two molecules after they have been docked. Most commonly one of the molecules is a small organic compound such as a drug and the second is the drug's biological target such as a protein receptor.

Table 2. 22 specific cases from CCDC/ASTEX validation test

			IBM x3650 M2		IBM BladeCenter HS22	
Test ID	ligand byte	Fitness	Bestranking time,s	Total run time,s	Bestranking time,s	Total run time,s
1a0q	982	88.82	34.18	37.30	32.14	36.23
1a1b	1430	100.51	196.37	203.20	194.80	198.12
1a1e	1430	99.40	184.43	191.06	182.20	190.47
1a4g	1073	91.52	36.02	40.46	32.10	35.38
1a4k	985	65.61	116.88	122.45	113.55	117.47
1a4q	1074	102.89	70.83	77.14	65.32	67.17
1a6w	807	60.22	21.53	22.55	20.46	21.57
1a07	1432	53.44	173.49	176.88	170.21	174.22
1a9u	1163	64.29	30.89	35.67	28.23	33.76
1a28	988	68.03	34.89	36.05	32.90	35.54
1a42	893	77.93	67.90	70.02	62.56	66.89
1aaq	545	94.53	259.27	263.46	249.18	253.38

			IBM x3650 M2		IBM BladeCenter HS22	
Test ID	ligand byte	Fitness	Bestranking time,s	Total run time,s	Bestranking time,s	Total run time,s
1abe	542	54.22	20.24	41.16	18.31	40.03
1abe	542	57.69	20.19	41.16	18.15	40.03
1abf	542	58.47	23.36	47.62	21.39	45.62
1abf	542	55.19	23.45	47.62	21.59	45.62
1acj	537	52.84	20.50	21.45	17.85	20.58
1acl	634	75.33	99.15	100.15	96.23	99.37
1acm	901	96.64	22.05	23.03	20.37	23.01
1aco	634	86.49	17.51	17.97	13.09	15.85
1aec	812	63.00	49.23	50.01	45.28	48.68
1aha	545	47.15	13.47	14.08	11.42	13.27
1ai5	988	47.59	15.93	16.83	13.25	15.49
1aj7	988	82.48	37.91	40.13	33.06	38.86

5 Discussion

Comparisons suggest that the best algorithm for docking is probably a hybrid of various types of algorithm encompassing novel search and scoring strategies. The most useful docking method will not only perform well, but will be easy to use and parametrise, and sufficiently adaptable such that different functionality may be selected, depending on the number of structures to be docked, the available computational resources, and the complexity of the problem. If the parameters cannot be generated quickly then although the algorithm may be computationally efficient, from a practical point of view it is limited. Conversely, a rapid scoring function may not necessarily be able to model some specific interactions. Moreover, although current docking methods show great promise, fast and accurate discrimination between different ligands based on binding affinity, once the binding mode is generated, is still a significant problem.

Acknowledgment

This research was supported by project „Bioinformatic research: protein folding, docking and predicting biological activity" NSF I02/16 12.12.14 funding by National Science Fund - Ministry of Education and Science of Bulgaria.

References

1. Halperin I, Ma B, Wolfson H, Nussinov R.Proteins **47** (4): 409–443 (June 2002).
2. Mustard D, Ritchie DW. Proteins **60** (2): 269–274 (August 2005).
3. Shoichet BK, Stroud RM, Santi DV, Kuntz ID, Perry KM. Science **259** (5100): 1445–50 (March 1993).
4. McGann MR, Almond HR, Nicholls A, Grant JA, Brown FK. Biopolymers **68** (1): 76–90 (January 2003).
5. Friesner RA, Banks JL, Murphy RB, Halgren TA, Klicic JJ, Mainz DT, Repasky MP, Knoll EH, Shelley M, Perry JK, Shaw DE, Francis P, Shenkin PS. J. Med. Chem. **47** (7): 1739–1749 (March 2004).
6. Jain AN. J. Med. Chem. **46** (4): 499–511 (February 2003).
7. Zsoldos Z, Reid D, Simon A, Sadjad SB, Johnson AP. J. Mol. Graph. Model. **26** (1): 198–212 (July 2007).
8. Jones G, Willett P, Glen RC, Leach AR, Taylor R. J. Mol. Biol. **267** (3): 727–748 (April 1997).

Adaptive Collaborative Quantum-Inspired Evolutionary Algorithm for Global Numerical Functions

Zhou Liang[1,*], *Shao* Ming[2], and *Ma* Chengqian[3]

[1]School of Information Management, Shanghai Lixin University of Accounting and Finance, Shanghai, China
[2]School of Management, Shanghai University of Engineering Science, Shanghai, China
[3]The college of Information, Mechnical and Electrical Engineering, Shanghai Normal University, Shanghai, China

Abstract. – A novel adaptive collaborative quantum-inspired evolutionary algorithm (ACQEA) is proposed by combining the collaborative evolution and adaptive mutation mechanism together in this paper. In ACQEA, the whole population will be divided into multi sub-populations which can complete the evolution independently, and then the collaborative evolution mechanism is used to make these multi sub-populations full exchange their evolution information in operation process. In addition, the adaptive mutation and update strategies are implemented in order to give ACQEA the power to explore its search space on the basis of characteristic information of the elite individual and corresponding population diversity. Finally, the proposed ACQEA is compared with existing quantum evolution algorithm (QEA) in solving global numerical functions and the experiments results verify that the advantages of ACQEA on convergence rate and searching accuracy.

1 Introduction

Quantum-inspired evolutionary algorithm (QEA) is proposed by combining quantum computing theory with evolutionary theory. It utilizes the concepts of quantum bit, superposition of states and collapse of states and it applies quantum rotation gate to guiding the searching direction into the optimal area and accelerating the algorithm's convergence speed [1]. Thus QEA have a number of computational advantages compared with the traditional optimization algorithms. QEA is still easy to fall into local optimum, especially for complex global optimization problems [2]. Recently, numerous literatures about improved QEA have been reported which focus on making its performance better, and these works can be summarized as following three aspects.

(1) The new encoding methods are proposed on QEA to improve the global search ability. Li Panchi et al. used Bloch coordinates comprise gene chain to represent quantum individuals, which can accelerate the convergence process [3]; Zhao Shuanfeng et al. proposed a real-coded chaotic quantum genetic algorithm based on the chaotic and coherent characters of Q-bits [4]; Zhang Lei et al. adopt phase angle and its corresponding real mapping to construct quantum individual with probabilistic representation, which can represent all the feasible solution probabilistically [5]; Jianyong Liu et al. had thought conventional quantum genetic algorithm (QGA) could be used to solve combinatorial optimization problems well than real-valued problems, so they proposed Real-coded QGA [6].

(2) New evolutionary strategies are adopted into QEA to improve its performance. Jianyong Liu et. al. combine quantum coding, immune clone and niche mechanism together to solve multi-modal function optimization more effectively [7]; Tzyy-Chyang Lu proposed a region-based QEA, the algorithm adopted quantum bit representation for the selection of regions; Its search process uses a region-by-region exploration in the beginning and, as the candidate regions are identified, a randomized search in good regions is employed for exploitation [8]; Qian Jie et al. proposed a statistical learning QEA. Experiments had shown that this new design could avoid premature convergence [9].

(3)The hybrid approaches are proposed which combines the traditional version with other optimization algorithms. Yu Yang et al. combined QEA with PSO and proposed two hybrid QEAs, one is to embed the evolutionary equation of PSO into QEA, while the other one is applied the quantum chromosomes of QEA to binary PSO [10]; Li Zhiyong et al. proposed a novel quantum bat algorithm, which is based on quantum evolution algorithm and bat algorithm [11]; Wu Shun-Chi et al. proposed a two-stage scheme that combines quantum evolutionary algorithm and tabu search, simultaneously possessed the capabilities of exploration and exploitation and could be effectively employed for LP design with less computational load [12].

In addition, co-evolutionary algorithms (CEAs), as a new evolutionary algorithm framework, focus on the idea of mutual influence or co-ordination among different species to ensure algorithms have better search accuracy and search efficiency [13]. In recent years, some researchers began to combine co-evolutionary mechanism with QEA and then proposed some improved QEAs. Gu et al. proposed a novel competitive co-evolutionary quantum genetic algorithm which included

* Corresponding author: hzhou0168@yahoo.com

three new strategies named as competitive hunter, cooperative surviving and the big fish eating small fish [14]. Xiong, Gui et al. proposed a double population co-evolution algorithm based on the quantum evolution algorithm and difference evolution algorithm. In their algorithm, one population is used for global searching; the other one is for partial searching [15]. Zhang proposed an elite collaborative quantum evolution algorithm. Two sub-populations were selected randomly. Subsequently, the corresponding elite individuals from the selected two sub-populations would construct a mutually beneficial relationship in order to implement co-evolutionary operation, and guide the individual populations evolve toward the optimal solution [16].

In this paper, the co-evolutionary mechanisms are introduced into QEA in order to propose Adaptive Collaborative Quantum-inspired Evolutionary Algorithm (ACQEA). This new ACQEA will fully extract the characteristic information of every sub-population in order to lead the algorithm has a better performance. The rest of this paper is organized as following. The preliminary knowledge of co-evolutionary algorithm and quantum evolution algorithm is explained in Section 2; section 3 gives the details of the proposed algorithm; section 4 shows the experimental simulation using the novel algorithm; finally, conclusions are drawn in section 5.

2 Preliminary knowledge

2.1 Co-evolutionary algorithm

Now there are two structural forms for CEAs to implement evolution. One is Competitive Co-evolutionary Algorithm (Com-CEA) and the other one is Cooperative Co-evolutionary Algorithm (Coo-CEA) [17][18]. In Coo-CEA, the fitness of individual is not evaluated separately. It is firstly in accordance with the prior knowledge of solving problem; subsequently, it combines the "representatives" individuals from the other sub-populations to construct a solution vector of solving problem and then to evaluate the fitness of individuals. This process is called "cooperation"; the cooperative co-evolution mechanism also provides a generalized evolution framework to solve the high-dimensional or complex optimization problems [19].

As for Com-CEA, the individual fitness in a sub-population is determined by the competition results which include a series of competition with the individuals in other sub-population. At this time, these two sub-populations always take the role of "Host" population and "Parasite". The evolutionary operations will use these populations to produce new sub-populations. With the development of the evolution, these different sub-populations will show different genetic characteristics.

2.2 Quantum evolution algorithm

Quantum evolution Algorithm (QEA) has advantages of quantum computing and evolutionary computation. It uses quantum bit (Q-bit) as a probabilistic representation to improve the diversity of population. It also uses quantum rotation gate (Q-gate) as the update operation, which can guide the search direction of individual to the optimal area, and increase the algorithm's convergence speed. In the traditional QEA, a Q-bit may be in the "1" state, "0" state, or any superposition of these two states. It can be represented as following:

$$|\Psi> = \alpha|0> + \beta|1> \quad (2\text{-}1)$$

Where α and β are complex numbers that specify the corresponding probability amplitudes of the states "0" and "1". Normalization of the two states to unity always guarantees:

$$|\alpha|^2 + |\beta|^2 = 1 \quad (2\text{-}2)$$

So a Q-bit in QEA can be defined with a pair of numbers: $[\alpha\ \beta]^T$.

In general, a quantum individual is defined as a string of Q-bits, $q_i = \begin{bmatrix} \alpha_{i1} & \alpha_{i2} & \cdots & \alpha_{im} \\ \beta_{i1} & \beta_{i2} & \cdots & \beta_{im} \end{bmatrix}$; the population $Q(t) = \{q_1, q_2, \dots, q_n\}$, where i is the size of population, and m is the number of Q-bits. Since the Q-bits representation is able to express as a linear superposition of states probabilistically, evolutionary computing with the Q-bit representation is often has a better performance than traditional approaches.

The main evolutionary way of QEA is realized by Q-gate, which transfer the probability of Q-bits converge either to "1" (or "0"). Q-gate can be designed according to the actual problem. In this paper, we use the common one as following.

$$U(\theta_i) = \begin{bmatrix} \cos(\theta_i) & -\sin(\theta_i) \\ \sin(\theta_i) & \cos(\theta_i) \end{bmatrix} \quad (2\text{-}3)$$

The update process of Q-gate and its related adjustment strategy is:

$$\begin{bmatrix} \alpha_i' \\ \beta_i' \end{bmatrix} = U(\theta_i) \begin{bmatrix} \alpha_i \\ \beta_i \end{bmatrix} \quad (2\text{-}4)$$

$$\theta_i = s(\alpha_i, \beta_i)\delta \quad (2\text{-}5)$$

Where θ_i is the rotation angle whose size and sign is definited by adjustment strategy in literature [20]. By using Q-gates, the quantum individuals can be guided toward the better solutions throughout the iteration process.

3 Adaptive collaborative quantum evolut ionary algorithm

This section proposes adaptive collaborative quantum evolutionary algorithm. Firstly, the initial population will be divided into multiple sub-populations; then the following new evolutionary strategies are introduced into these multiple sub-populations in order to improve its global optimization ability.

3.1 Elite individuals

Elite individual generally corresponds to the individual with a better fitness and contains corresponding evolutionary characteristic information. Maintaining best Q-bit individuals as elite individual in every generation can avoid the possibility of losing high quality individuals. At the same point, it is even more powerful if the elitism is further strengthened and the solutions are spread out by quantum mechanism [21]. In ACQEA, every sub-population will produce an elite individual which actually contains the characteristics information of current sub-population in the implementation process of algorithm. Thus, these elite individuals can be collected as guiding individual. Accompanied by the evolution of the whole population, these guiding elites will be applied to help the poor performance sub-population and constantly updated.

3.2 Definition of population diversity

Current experience has shown that the population diversity is very important in giving the algorithm the power to explore the search space and not get trapped in local optima [22]. Normally, if the diversity of sub-populations is not identified at the end of iteration, it would be gradually reduced in the later iterative process of the algorithm. This could result in a gene decrease for effective individuals, and affect the optimization process of the algorithm and finally cause a premature convergence. Therefore, it is necessary to design a mechanism to maintain the diversity of the population.

In this section, a modified indicator of degree of population diversity is introduced to measure the diversity of sub-populations based on the reference [23]. Here the modified indicator includes the characteristic information of every individual in each sub-population and also is a parameter involved in the evolution of the population.

Definition 3.1 When the Q-bits are applied to optimization problems, they will converge to the corresponding binary encoding space $\{0,1\}^L$, where L is the length of binary code, the population size is n. A group of individuals in this population can be described: $Q = \{q_1, q_2, \dots, q_n\}$, where $q_j = \{q_{1j}, q_{2j}, \dots, q_{Lj}\}$, and $j = 1,2,\dots,n$. The degree of population diversity is defined as the following formula:

$$\mathrm{div}(Q) = \exp\left(-\frac{\left|\sum_{l=1}^{L}\sum_{j=1}^{n} q_{lj} - \sum_{l=1}^{L}\sum_{j=1}^{n}(1-q_{lj})\right|}{\mathrm{scale}}\right) \quad (3\text{-}1)$$

where scale is the size of individuals in population, and $\mathrm{scale} = L \times n$, $\sum_{j=1}^{n} q_{lj}$ represent the numbers of binary code "1" of all the individuals in their lth bit in the population; while $\sum_{j=1}^{n}(1-q_{lj})$ represent the numbers of binary code "0" of all the individuals in their lth bit in the population. The result of calculating $\left|\sum_{l=1}^{L}\sum_{j=1}^{n} q_{lj} - \sum_{l=1}^{L}\sum_{j=1}^{n}(1-q_{lj})\right|$ indicates that the distribution of binary code "0" and "1" of all the individuals in the population. In each iteration, the greater the result, the more unbalanced distribution between the binary code "1" and "0", which also means the population diversity is in reducing. Finally, when the result of div(Q) becomes smaller, it shows the diversity of the population is getting worse. As mentioned in above formula (3-1), when the ACQEA divides the population into many sub-populations, every sub-population will has its own div(Sub_Q) as an indicator to measure the sub-population diversity during the entire evolutionary process.

3.3 Adaptive mutation and update mechanism

Mutation operator can produce a random disturbance for the candidate solution in order to get new generations. In general, the efficiency of evolutionary algorithm relies strongly on the performance of mutation operation [24]. The ACQEA in this section will implement mutation operation for the worst sub-population with smallest div(Q) in order to improve its diversity. Here should be noted that the change of div(Q) is related to the probability of mutation. Therefore, when the div(Q) becomes smaller, it means the diversity of population is getting worse, while the mutation probability of the sub-population will increase. Here the diversity of population and global searching abilities would be improved by adaptively changing of the mutation probability.

With this process, if the original value of the Q-bit is '0', then it will turn into '1' after this mutation operation; the other way around is '1', then it will turn into '0' after mutation operation. When the QEA used Q-gate to update these sub-populations, the best individual with in the elite library will be used to guide and update the following two sub-populations instead of the whole. One is the sub-population which is corresponding to the worst individual in elite library, and the other is the worst diversity of the population in order to ensure the poor sub-populations have a better optimization performance. The remaining sub-populations will be updated by their own best individuals.

As a whole, the mutation and update strategy is actually equivalent to build up an information exchange mechanism between the best sub-population and the worst sub-population, at the same time, this mechanism also ensure some of the sub-populations can maintain their independent evolutions.

4 Simulation tests

In order to illustrate the effectiveness and performance of the proposed ACQEA, we compare it with the traditional QEA [25] by using the following typical global numerical functions with diverse complexities from virtual library of Simulation Experiments in Simon Fraser University. These functions listed below are some of the common functions used for testing optimization algorithms, and all of them have many local minima points. The maximum number of the iteration of these two algorithms is 100, and the initial population size is 200. The population of ACQEA is divided into four sub-populations with size of 50 each.

4.1 Description of the test functions

These test functions are listed as following:

Function 1: Ackley Function

$$f(x) = -a\exp\left(-b\sqrt{\frac{1}{d}\sum_{i=1}^{d} x_i^2}\right) - \exp\left(\frac{1}{d}\sum_{i=1}^{d}\cos(cx_i)\right) + a + \exp(1)$$

Where d is the dimensions; The Ackley function is widely used for testing optimization algorithms. In its two-dimensional form in the plot above, it is characterized by a nearly flat outer region, and a large hole at the centre. The function poses a risk for optimization algorithms, particularly hillclimbing algorithms, to be trapped in one of its many local minima. Recommended variable values are: a = 20, b = 0.2 and c = 2π. Its corresponding input domain is on $x_i \in [-32.768, 32.768]$, for all $i = 1,2,\ldots,d$. Its global minimum $f(x^*) = 0$ at $x^* = (0,0,\ldots,0)$.

Function 2: Bukin Function N. 6

$$f(x) = 100\sqrt{|x_2 - 0.01x_1^2|} + 0.01|x_1 + 10|$$

The sixth Bukin function has many local minima, all of which lie in a ridge; and it is usually evaluated on the rectangle $x_1 \in [-15,-5]$, $x_2 \in [-3,3]$. Its global minimum $f(x^*) = 0$ at $x^* = (-10,1)$.

Function 3: Cross-in-Tray Function

$$f(x) = -0.0001\left(\left|\sin(x_1)\sin(x_2)\exp\left(\left|100 - \frac{\sqrt{x_1^2 + x_2^2}}{\pi}\right|\right)\right| + 1\right)^{0.1}$$

The Cross-in-Tray function has multiple global minima. It is shown here with a smaller domain in the second plot, so that its characteristic "cross" will be visible. The function is usually evaluated on the square $x_i \in [-10,10]$, for all $i = 1,2$. Its global minimum $f(x^*) = -2.06261$ at:

$$x^* = (1.3491,-1.3491),(1.3491,1.3491),(-1.3491,1.3491),(-1.3491,-1.3491).$$

Function 4: Drop-Wave Function

$$f(x) = -\frac{1 + \cos(12\sqrt{x_1^2 + x_2^2})}{0.5(x_1^2 + x_2^2) + 2}$$

The Drop-Wave function is multimodal and highly complex. The second plot above shows the function on a smaller input domain, to illustrate its characteristic features. The function is usually evaluated on the square $x_i \in [-5.12, 5.12]$ for all $i = 1,2$. Its global minimum $f(x^*) = -1$ at: $x^* = (0,0)$.

Function 5: Eggholder Function

$$f(x) = -(x_2 + 47)\sin\left(\sqrt{\left|x_2 + \frac{x_1}{2} + 47\right|}\right) - x_1\sin\left(\sqrt{|x_1 - (x_2 + 47)|}\right)$$

The Eggholder function is a difficult function to optimize, because of the large number of local minima. The function is usually evaluated on the square $x_i \in [-512,512]$, for all $i = 1,2$. Its global minimum $f(x^*) = -959.6407$ at: $x^* = (512, 404.2319)$.

All the above numerical optimization benchmark functions have many local minima around the global minimum, therefore they can be used to test the global search ability of the algorithm and its ability to escape from local optima. Notices that the above numerical optimization benchmark functions are used to solve the minimum, and we will reverse these optimization functions into solving the maximum in order to handling uniformity. Each algorithm runs 10 times independently. The statistical results including the best optimal result, the average result, the worst results, and the standard deviations are described as following table.

4.2 Simulation results

Table 1 shows the overall statistical results of QEA and ACQEA for the above functions. The statistical results indicate that the ACQEA is superior to QEA from the view of quality of solutions when these two optimization algorithms perform the same functions from table 1. The results show that the ACQEA is more efficient in finding the global optimal solution and robustness than QEA. The standard deviation indicates that ACQEA has a good stability and widespread adaptability.

Table 1. Statistical Results of QEA and ACQEA for Optimization Functions (with 10 experiments)

Test Functions	QEA			
	Best	Worst	Average	Standard Deviation
$f_1(x)$	-1.2505e-004	-1.1839e-003	-4.8783e-004	3.69073e-004
$f_2(x)$	-0.036344	-0.12651	-0.081246	0.026744
$f_3(x)$	2.0626	2.0626	2.0626	0
$f_4(x)$	1	0.93625	0.98725	0.0255
$f_5(x)$	959.6407	955.1846	957.9439	1.92930
Test Functions	**ACQEA**			
	Best	Worst	Average	Standard Deviation
$f_1(x)$	-1.2505e-004	-9.7945e-004	-3.41813e-004	2.95544e-004
$f_2(x)$	-0.021733	-0.10834	-0.058894	0.028338
$f_3(x)$	2.0626	2.0626	2.0626	0
$f_4(x)$	1	0.99999	0.999999	2.999999e-006
$f_5(x)$	959.6407	955.3586	958.1830	1.76522

In addition, we take the evolutionary results for function 5 based on ACQEA for example, although each sub-population is evolving independently, these four sub-populations will become synchronized toward convergence in the later iterative process. Here for function 5, their corresponding sub-populations can almost converge to their optimal solutions throughout the iteration cycle at the same time. As for function 5 in Figure 1, the optimal solutions in sub-population 1, 2, 3,

4 are (512, 404.2079), (511.999, 404.2186), (511.999, 404.246), (512, 404.2323) respectively; and their corresponding optimal values are 959.64, 959.6372, 959.6371, 959.6407. The results also present the following three conclusions. Firstly, the new evolutionary strategies are effective for the adaptive collaborative quantum evolutionary algorithm. On the other hand, the elite individual in the constructing elite library just guide the selected sub-population instead of entire population toward the global optimum, which is enough to complete the optimize search and can improve the operating efficiency of the algorithm to a certain degree. Thirdly, Our ACQEA can use smaller populations to complete complex optimization problems with the help of co-evolutionary mechanism and performs better than traditional QEA regardless the complexity of test functions.

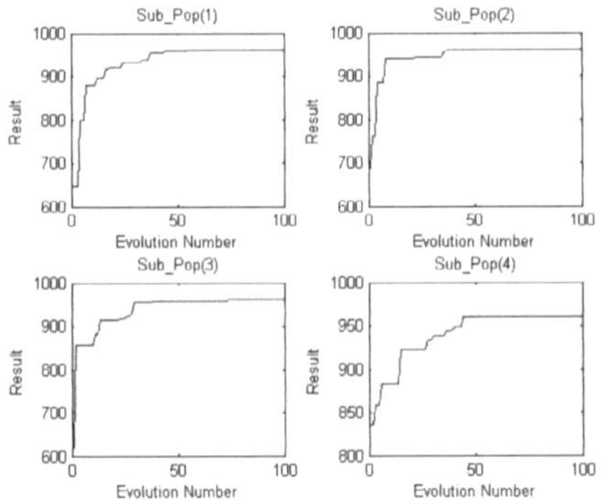

Fig.1. Evolutionary results for Function 5 on the basis of ACQEA

5 Conclusions

An Adaptive Collaborative Quantum Evolutionary Algorithm is proposed in this paper. In ACQEA, the whole population is firstly divided into multiple sub-populations which complete the evolution process independently; and then makes full use of the characteristic information of each sub-population with the new evolutionary strategy to implement the evolution process. ACQEA is compared with existing QEA using the established test functions. The results of simulation experiments for ACQEA show that its convergence performance is superior to QEA; and ACQEA is more accurate and stable in finding the global value in the functions. In the future work, how to apply the algorithm to the more complex multi-objective optimization problems and adjust the parameters of this algorithm in order to ensure it can get the best performance need more research.

References

1. M. Shao, L. Zhou, Novel Quantum-Inspired Co-evolutionary Algorithm, *International Journal of Security and Its Applications*, **10**, 2 (2016)
2. Y.G. Cai, M.J. Zhang, H. Cai,Y. Zhang, Hybrid Chaotic Quantum Evolutionary Algorithm, *Systems Engineering – Theory & Practice*, **32**, 10 (2012)
3. P.C. Li, S.Y. Li. Quantum-inspired evolutionary algorithm for continuous space optimization based on Bloch coordinates of qubits, Neurocomputing, **72**, 1/2/3 (2008)
4. S.F. Zhao, G.H. Xu, T.F. Tao, L. Lin, Real-coded chaotic quantum-inspired genetic algorithm for training of fuzzy neural networks, *Computers and Mathematics with Applications*, **57**, 11-12 (2009)
5. L. Zhang ,Y.W. Fang , D.H. Mao , P.F. Yang, A new phase angle encoded quantum evolutionary algorithm, *Control and Decision*, **30**, 4 (2015)
6. J.Y. Liu, H.X. Wang, Y.Y. Sun, C.Q. Fu, J. Guo, Real-Coded Quantum-Inspired Genetic Algorithm-Based BP Neural Network Algorithm, *Mathematical Problems in Engineering*, **2015**, (2015)
7. J.Y. Liu, H.X. Wang, Y.Y. Sun, L. Li, Adaptive niche quantum-inspired immune clonal algorithm, *Natural Computing*, **15**, 2 (2016)
8. T.C. Lu,J.C. Juang, A region-based quantum evo lutionary algorithm for global numerical optimiz ation, *Journal of Computational and Applied Ma thematics*, **239** (2013)
9. J. Qian, J.G. Zheng, A Quantum Evolutionary A lgorithm Based on Population Statistical Learnin g, *Journal of Xi'an Jiaotong University*, **46**, 2 (2 012)
10. Y.Yu, Z.F. Yin,Y,F. Tian, Hybrid Quantum Evo lutionary Algorithms and its Application, *Comp uter Engineering and Applications*, **42**, 28 (2006)
11. Z.Y. Li, L. Ma, H.Z. Zhang, Quantum Bat Algorithm for Function Optimization, *Journal of System and Management*, **23**, 5 (2014)
12. S.C. Wu, T.H. Chan, Hsieh, M. Shao,C. Lin, Quantum evolutionary algorithm and tabu search in pressurized water reactor loading pattern design, *Annals Of Nuclear Energy*, **94**, (2016)
13. X.Y. Liu, J. Liu, G.R. Zhao, J. Ma, Y.C. Wang, Study on cooperation evolution model, Computer Engineering and Applications, **48**, 8 (2012)
14. J.W. Gu, M.Z. Gu, C.W. Cao, X.S. Gu, A novel competitive co-evolutionary quantum genetic algorithm for stochastic job shop scheduling problem, *Computers & Operations Research*, **37**, 5 (2010)
15. F.Q. Xiong, W.H. Gui, C.H. Yang, Y.G. Li, A double population co-evolution algorithm for

process of zinc hydrometallurgy, *Control and Decision*, **28**, 4 (2013).
16. W.F. Zhang, Elite Collaborative Quantum Evolution Algorithm for Combinatorial Optimization Problem, *Journal of Hubei University of Automotive Technology*, **27**, 3 (2013)
17. M.Q. Li., D. Lin., J.S. Kou, Dynamics of Fitness Sharing Evolutionary Algorithms for Coevolution of Multiple Species, *Applied Soft Computing*, **10**, 3 (2010)
18. X.Y. Sun, D.W. Cong, Varying population size cooperative coevolutioanry genetic algorithm and its application in optimization, *Control and Decision*, **19**,12 (2004)
19. X.X. Hu,Y.L. Yin, Cooperative Co-Evolutionary Cuckoo Search Algorithm for Continuous Function Optimization Problems, *Pattern recognition and Artificial Intelligence*, **26**, 11 (2013)
20. L.F. Zhang, X.P. Zhang, Network traffic prediction based on BP neural networks optimized by quantum genetic algorithm, *Computer Engineering and Science*, **38**, 1 (2016)
21. B. Jerzy, D. Grzegorz, M. Jan, Multi-criterion, Evolutionary and Quantum Decision Making in Complex Systems, *Recent Advances in Applied & Biomedical Informatics and Computational Engineering in Systems Applications* (WSEAS PRESS, 2011)
22. H. Mohammad. N.Tayarani, M. R. Akbarzadeh-T, Improvement of the performance of the Quantum-inspired Evolutionary Algorithms: structures, population, operators, *Evolutionary Intelligence*, **7**, 4 (2014)
23. Q. Liu, X.Y. Wang, Q.M. Fu, Y.G. Zhang, X.F. Zhang, Double Elite Coevolutionary Genetic Algorithm, *Journal of Software*, **23**, 4 (2012)
24. S.H. Wen, J.H. Zheng, M.Q. Li, Comparison and research of mutation operators in multi-objective evolutionary algorithm, *Computer Engineering and Applications*, **45**, 2 (2009)
25. J. Qian, J.G. Zheng, C.Q. Zhang, X. Wang, R.X. Yan, Reviews of current studying progress on quantum evolutionary computation, *Control and Decision* , **26**, 3 (2011)

ITM Web of Conferences **16**, 03001 (2018) https://doi.org/10.1051/itmconf/20181603001
AMCSE 2017

The Study of Permanent Magnets Synchronous Machine (PMSM) of the Autonomous Electric Power Supply System (ASE), compatible with the Concept of a More Electric Aircraft (MEA)

Lucjan Setlak[1*], *Rafał* Kowalik[1]

[1]Department of Avionics and Control Systems, Polish Air Force Academy, ul. Dywizjonu 303 nr 35, 08-521 Deblin

Abstract. Based on the analysis and mathematical models of synchronous electric machines (motor/generator), basing on permanent magnets, presented in this paper, the main importance of alternator AC power sources in the form of starter/generator (for conventional aircraft) and in the form of integrated unit starter (motor)/AC synchronous generator S/G AC (with respect to advanced aircraft concept in terms of more electric aircraft) was highlighted. Additionally, through the analysis and selected simulations of the on-board autonomous power supply system of the modern aircrafts, sources of electrical energy (synchronous motor/generator, integrated unit starter/AC generator) were located in board autonomic power system ASE (EPS, PES). Main components of this system are the electro-energetic power system EPS and the energo-electronic power system PES. In addition, the analysis and exemplary simulations of main electricity sources based on mathematical models have contributed to highlighting the main practical applications in accordance with the concept of MEA.

1 Introduction

Currently, in modern aircrafts of the main airline companies, both civil aviation (Airbus, Boeing, etc.) and military (Lockheed Martin), can be seen continuous development in terms of electrical machines [1], [2], which occurs due to dynamic technological progress of analogous fields of science, especially the innovative power electronics systems PES, equipment and advanced materials, and what goes with it – the new integrated avionics [3]. Additionally, gradually increasing dynamics of changes in the board power supply in modern aviation can be seen, both in the field of civil aircrafts (Airbus, Boeing) for airplanes (A-380 and A-350XWB, B-787), as well as for military aviation (Lockheed Martin) in terms of aircrafts JSF (Joint Strike Fighter) F-35 and F-22 Raptor. Changes in on-board power supply in modern aviation have a key impact on the rapidly developing concept of a more electric plane, including in particular the dominant role of advanced systems in the production or processing of AC and their main components (rectifiers, transducers, etc.), as well as the EPS system in the area of direct current (DC generator or generators, accumulators as emergency sources, electric energy storage tanks, fuel cells, etc.). When making an initial analysis of the dominant role of on-board alternating current and direct current DC power supply, it should be noted that the on-board power supply is usually based on main sources of synchronous electrical machinery (motor, generator). In the case of "classic" aircraft, these are generators or integrated unit DC motor/DC generator and generators of increased constant frequency (CF) 400 Hz of alternating current (AC). In turn, in the case of "More Electric Aircraft", they are the source generators or integrated units of variable frequency (VF) generators, ranging from 390-780 [Hz].

In view of the above, the on-board power supply of a modern aircraft in line with the new trend of the "More Electric Aircraft" concept may be based on a PMSM synchronous generator or an integrated unit starter AS/G VF (Alternating/Starter Variable Frequency Generator) [4], [5].

Based on the above, modern aircraft use primarily 3-phase synchronous machines (e.g. generators) as a source of high-power AC electricity. In addition, it is noted that such PMSM electric machines operate most effectively during a generative range of operation, i.e. in terms of power produced. This does not mean, however, that PMSM synchronous electric machines can not operate on a motor (starter) operating range, that is, in the start-up range. Later the paper discusses and presents mathematical models of synchronous PMSM electric machines on the example of a synchronous motor for both serial and parallel connections in the field of EPS power supply, which is the main ASE power supply system implemented on modern aircraft.

* Corresponding author: l.setlak@wsosp.pl, r.kowalik@wsosp.pl

 https://doi.org/10.1051/itmconf/20181603001

2 Analysis and mathematical model of ASE system compatible with the trend of a More Electric Aircraft

Analysis and creating a mathematical model in the electro-energetic power system of a more electric aircraft was performed using an electric machine (motor/generator) based on permanent magnets PMSM. Proceeding with the mathematical description (model) of physical phenomena occurring in the electrical machine PMSM, responsible for the generation and transmission of electricity needed to power the main board equipment (aerial flight instruments, aircraft electrical equipment), implemented on a modern airplane, at the beginning you should consider the above issues with regard to the interaction of the magnetic field formed between the stator and the rotor of the electrical machine.

At the next stage of the analysis, examining the physical phenomena occurring in the PMSM electrical machine, particular attention was paid to the process of starting the engine. During the induction of the stator winding with alternating voltage, a magnetic field is generated in the PMSM electrical machine. The speed at which the rotor rotates in a unit of time is called the rotational speed of the synchronous motor. In the initial phase it was assumed that the engine was stationary. In order to start the synchronous electrical machine PMSM, you should rotate the machine rotor at a speed close to or equal to the engine rotational speed. At a later stage of the start-up process, the rotor unit is driven by the energy supplied by the propulsion system of the aircraft. If the synchronous speed is reached by the electric machine, the DC power is supplied to the rotor winding and the motor starts. Once the engine is started, a magnetic field is generated at the same time. Then, at some point, the magnetic field created by the rotor enters the stator and the engine is running at a certain rotational speed. It should be noted that in the motor (starter) operating range, the PMSM electrical machine rotor rotates at a synchronous speed. In addition, the engine operating range allows the performance in a single coordinate system respective moments of rotation of the four permanent magnets fixedly connected to the stator.

The initial assumptions are that the magnetic field is generated in the right direction and that the direction of the magnetic field of the rotor coincides with the direction of the magnetic field appearing on the stator [6], [7]. Then, using Ampere's law relating the electromotive force (emf) and the distribution on the individual permanent magnets shown in Fig. 1, one can write a mathematical expression defining the electromotive force induced in the coil as

$$F = Bl_0 i \tag{1}$$

where B is the magnetic induction of the stator; l_o is the width of the coil, and i determines the alternating current flowing through the coil.

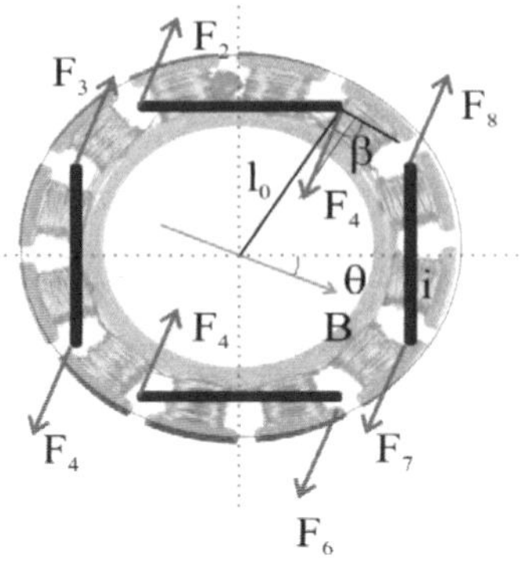

Fig. 1. Location of permanent magnets before starting the electric synchronous machine PMSM

Defining the value of the electromagnetic torque of a 4-pole synchronous motor, the PMSM rotor can be used as a series connection of coils and resistors [8]. Resistance elements are responsible for creating electromagnetic forces F_k, $k = 1, \ldots, 8$. Therefore, present in the analyzed electrical machine PMSM forces F_1 and F_2 act only on the first layers of the coil and are determined by (Fig. 2)

$$\begin{aligned} F_{1pr} &= F_1 \cos\beta_1 = Bl_0 i \cos(\theta + \alpha) \\ F_{2pr} &= F_2 \cos\beta_2 = Bl_0 i \cos(\alpha - \theta) \end{aligned} \tag{2}$$

where β_1 is the angle between the force vector F_1 and the perpendicular line defining the position of the permanent magnets relative to the motion of the magnetic field formed in the stator; β_2 the angle between the force vector F_2 and the perpendicular line, defining the position of the permanent magnets relative to the motion of the magnetic field formed in the stator; α defines the angle between the magnetic field vector and the magnetic field of the synchronous motor coil for the first coil layer; and θ the angle between the plane perpendicular to the direction of the vector defining the stator magnetic field and the plane of the magnetic field of the coil in the first layers of the coil.

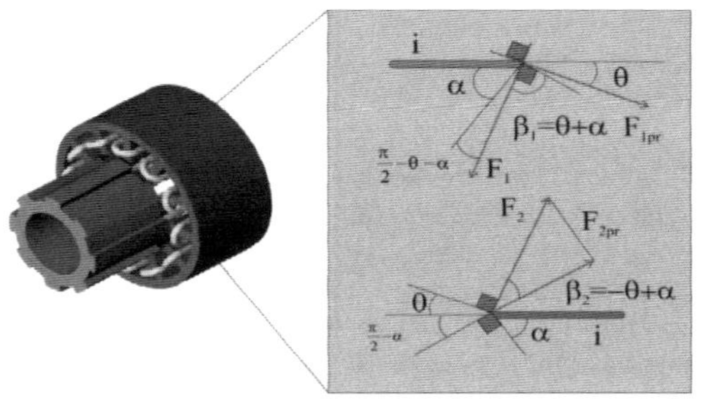

Fig. 2. Arrangement of vectors defining the position of permanent magnets

Next, taking into account the number of windings in the coil and the direction of rotation of the rotor, you can determine the mathematical relations that determine the torque acting on the coil. So the equations are as follows [9], [10]

$$M_1 = nl\left(F_{1pr} + F_{2pr}\right) = nlBl_0 i[\cos(\theta+\alpha) + \cos(\alpha-\theta)] = n(2ll_0\cos\alpha)Bi\cos\theta = nSBi\cos\theta \quad (3)$$

where n is the number of poles pairs of magnets in the induction motor; l is the length of the vector between individual permanent magnets, and S is the area of permanent magnets.

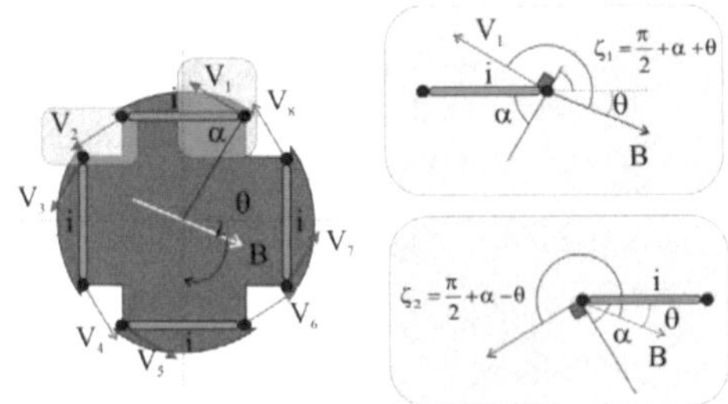

Fig. 3. The structure of a quadratic rotor at the parallel connection and the orientation of the magnetic flux vectors along with the definition of the angles ξ_1, ξ_2.

The electromagnetic moments in the other windings are determined in a similar way. Therefore

$$M_2 = nSBi\cos\left(\theta+\frac{\pi}{2}\right) = -nSBi\sin\theta \quad M_3 = nSBi\cos\theta \quad M_4 = -nSBi\sin\theta \quad (4)$$

Hence, the rotational electromagnetic torque of a synchronous machine with a four pole rotor is described by the equation

$$M_{em} = M_1 + M_2 + M_3 + M_4 = 2nSBi(\cos\theta - \sin\theta) = 2\sqrt{2}nSBi_3\sin\left(\theta+\frac{\pi}{2}\right) \quad (5)$$

In the case of a parallel connection of the resistive elements, the torque of the synchronous motor is defined as

$$M_{em} = nSB[\cos\theta(i_1+i_3) - \sin\theta(i_2+i_4)] \quad (6)$$

where i_1, i_2, i_3, i_4 are the currents flowing through the coils. Other sizes occurring in the equation have the same meaning as in earlier mathematical expressions.
Further mathematical considerations assume that all four coils have the same inductance. In this situation, the equation describing the dynamic phenomena occurring in the synchronous motor will take the form

$$J\frac{d^2\theta}{dt^2} = M_{em} - M_1 \quad (7)$$

where θ corresponds to the mechanical rotation angle of the rotor; M_{em} is the total electromagnetic torque with respect to systems that perform engine control functions. The next step is to determine the voltage equations for the substitute serial circuit defining a four-pole synchronous motor. Fig. 3 shows the replacement circuit diagram of the electrical rotor of the synchronous machine PMSM. Turning to the next stage of mathematical analysis, it is important to consider the system shown in Figure 2.

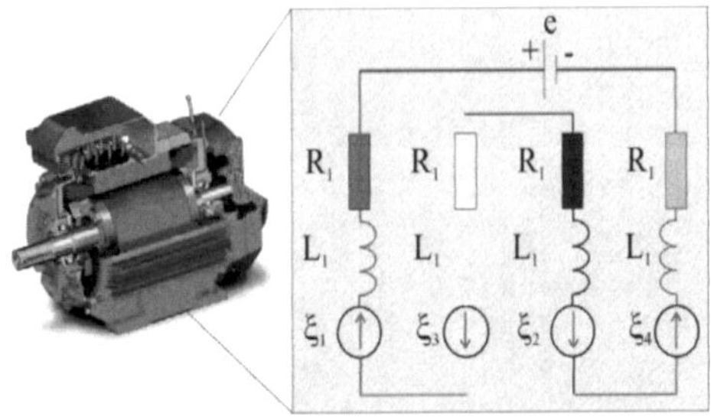

Fig. 4. Circuit diagram of a four-pole rotor circuit in serial connection

Thus, for each coil alternating current is defined in the same manner as in the previous considerations and taking into account Kirchhoff's second law to the closed circuit and taking the positive direction of the eyelet perimeter in layout of the analyzed substitute circuit arrangement according clockwise, the differential equation was defined [11], [12]

$$L\frac{di}{dt} + Ri = F_1 - F_2 + F_3 - F_4 + e \quad (8)$$

where R, L denote the resistance and inductance respectively in each coil; F_k determines the value of the electromotive force present in the k-coil as a result of the rotation of the rotor, and e is a voltage source. The directions of electromotive force are illustrated in Fig. 3. Thus, according to the law of electromagnetic induction, the electromotive force generated in the coil as a result of electromagnetic field radiation and rotor rotation is defined by the formula

$$F_1 = l_0Bv_1 sin\varphi_1 + l_0Bv_2 sin\varphi_2 \quad (9)$$

where v_1, v_2 represent respectively the velocity of change of the electromagnetic field relative to the coils, as illustrated in Fig. 4, and φ_1, φ_2 represent the angles between the velocity vector and the electromagnetic field vector. Such a solution is depicted in Fig. 5. In view of the above, the following was obtained

$$F_1 = -nl_0Bl\omega\left[\sin\left(\frac{\pi}{2}+\alpha+\theta\right) + \sin\left(\frac{\pi}{2}+\alpha-\theta\right)\right] = -n(2l_0l\cos\alpha)B\omega\cos\theta = -nSB\omega\cos\theta \quad (10)$$

where $\omega = \frac{d\theta}{dt}$. The electromotive forces on the remaining coils, as described by the following equations, were determined in the same way

$$F_2 = -nSB\omega\sin\theta \quad F_3 = -nSB\omega\cos\theta \quad F_4 = -nSB\omega\sin\theta \quad (11)$$

In contrast to a serial electrical circuit of the rotor, an equivalent circuit of parallel connection illustrated in Fig. 5, operates in closed feedback loop. Thus, by using the second Kirchhoff's law for each eyelet separately, choosing the positive direction of the eyelet circuit analyzed in a substitute circuit arrangement clockwise, as in the previous mathematical analysis.

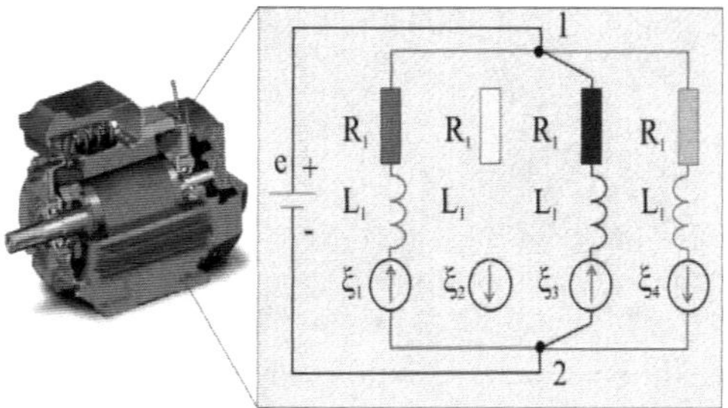

Fig. 5. Equivalent circuit diagram of a four pole rotor circuit connected in parallel

On this basis, the following differential equations were found to determine the currents in the circuit

$$\begin{aligned} L\frac{di_1}{dt} + Ri_1 &= F_1 + e = -nSB\omega\cos\theta + e \\ L\frac{di_2}{dt} + Ri_2 &= -F_2 + e = nSB\omega\sin\theta + e \\ L\frac{di_3}{dt} + Ri_3 &= F_3 + e = -nSB\omega\cos\theta + e \\ L\frac{di_4}{dt} + Ri_4 &= -F_4 + e = nSB\omega\sin\theta + e \end{aligned} \tag{12}$$

The values of the electromotive forces in the parallel circuit are determined identically as they have been before in equations (10) and (11). Thus, the dynamics of operation (action) of a 4-pole synchronous motor in the case of a serial connection can be described by the following differential equations

$$\begin{aligned} \frac{d\theta}{dt} &= \omega \\ J\frac{d\omega}{dt} &= 2\sqrt{2}SBi\sin\left(\frac{\pi}{4} - \theta\right) - M_J \\ L\frac{d^2i}{dt^2} + Ri &= 2\sqrt{2}SBi\sin\left(\frac{\pi}{4} - \theta\right) + e \end{aligned} \tag{13}$$

Accordingly, for parallel connection were obtained

$$\begin{aligned} \frac{d\theta}{dt} &= \omega \\ J\frac{d\omega}{dt} &= nSB[\cos\theta(i_1 + i_3) - \sin\theta(i_2 + i_4)] \\ L\frac{di_1}{dt} + Ri_1 &= -nSB\omega\cos\theta + e \\ L\frac{di_2}{dt} + Ri_2 &= nSB\omega\sin\theta + e \\ L\frac{di_3}{dt} + Ri_3 &= -nSB\omega\cos\theta + e \\ L\frac{di_4}{dt} + Ri_4 &= nSB\omega\sin\theta + e \end{aligned} \tag{14}$$

As mentioned earlier, most modern synchronous motors are used in advanced electric aircrafts systems (Airbus, Boeing, Lockheed Martin), they incorporate additional damping windings to reduce power losses in the initial phase of synchronous machine start-up. Damping windings in synchronous motors are mounted in the form of cages. Considering in the mathematical analysis of damping windings, the current equation assumes the following form

$$L_1\frac{dj_k}{dt} + R_1 j_k = -r_0 l_1 B\cos\left(\theta + \frac{2k\pi}{n_1}\right)\frac{d\theta}{dt}, \quad k = 1, \ldots, n_1 \tag{15}$$

In turn, the electromagnetic torque of the damping winding is described by the equation

$$M_{em\,ut} = r_0 l_1 B\sum_{k=1}^{n_1}\cos\left(\theta + \frac{2k\pi}{n_1}\right) \tag{16}$$

where n_1 is the number of poles; j_k current in the k – th damping winding; L_1 inductance of damping winding; R resistance of damping winding; r_0 and l_1 denote the radius and length of the single cage as part of the damping winding, respectively.

In practical applications, the electromagnetic fields of the damping winding and the rotor are presented as two independent, so in the case of a serial connection of the substitute circuit of a four pole synchronous motor, the mathematical model is as follows [13]

$$\begin{aligned} \frac{d\theta}{dt} &= \omega \\ J\frac{d\omega}{dt} &= 2\sqrt{2}SBi\sin\left(\frac{\pi}{4} - \theta\right) + r_0 l_1 B\sum_{k=1}^{n_1}\cos\left(\theta + \frac{2k\pi}{n_1}\right)j_k - M_J \\ L\frac{di}{dt} + Ri &= -2\sqrt{2}SBi\sin\left(\frac{\pi}{4} - \theta\right) + e \\ L\frac{dj_k}{dt} + Rj_k &= -r_0 l_1 B\cos\left(\theta + \frac{2k\pi}{n_1}\right)\frac{d\theta}{dt}, \quad k = 1, \ldots, n_1 \end{aligned} \tag{17}$$

And for parallel connection obtained

$$\begin{aligned} \frac{d\theta}{dt} &= \omega \\ J\frac{d\omega}{dt} &= nSB[\cos\theta(i_1 + i_3) - \sin\theta(i_2 + i_4)] + r_0 l_1 B\sum_{k=1}^{n_1}\cos\left(\theta + \frac{2k\pi}{n_1}\right)j_k - M_J \\ L\frac{di_1}{dt} + Ri_1 &= -nSB\omega\cos\theta + e \\ L\frac{di_2}{dt} + Ri_2 &= nSB\omega\sin\theta + e \\ L\frac{di_3}{dt} + Ri_3 &= -nSB\omega\cos\theta + e \\ L\frac{di_4}{dt} + Ri_4 &= nSB\omega\sin\theta + e \end{aligned} \tag{18}$$

$$L_1 \frac{dj_k}{dt} + R_1 j_k = -r_0 l_1 B \cos\left(\theta + \frac{2k\pi}{n_1}\right)\frac{d\theta}{dt},$$
$$k = 1, \dots, n_1$$

After making proper transformations, among others, by properly rearranging the size and adding additional parameters, obtained

$$\begin{gathered} \vartheta = -\theta - \frac{3\pi}{4} \\ s = -\omega \\ x = i + \frac{e}{R} \\ \mu = -\frac{2L_1}{n_1 r_0 l_1 B}\sum_{k=1}^{n_1} i_k \sin\left(\theta + \frac{2k\pi}{n_1}\right) \\ v = -\frac{2L_1}{n_1 r_0 l_1 B}\sum_{k=1}^{n_0} i_k \sin\left(\theta + \frac{2k\pi}{n_1}\right) \end{gathered} \tag{19}$$

$$z_k = \sum_{k=-\frac{n_1}{4}}^{\frac{n_1}{4}} i_{(k+j)} + i_k ctg\frac{\pi}{n_1}, \quad k = 3, \dots, n_1$$

As a result obtained

$$\begin{gathered} \frac{d\vartheta}{dt} = s \\ \frac{ds}{dt} = ax\sin\vartheta + bv - \delta_1(\vartheta) \\ \frac{d\mu}{dt} = -c_1\mu + vs\sin\vartheta \\ \frac{dx}{dt} = -c_1 x + ds \\ \frac{dv}{dt} = -c_1 v - s\mu - s \\ \frac{dz_k}{dt} = -c_1 z_k, \quad k = 3, \dots, n_1 \end{gathered} \tag{20}$$

where

$$\begin{gathered} a = 2\sqrt{2}\frac{nBS}{J}; \quad b = \frac{n_0(S_0B)^2}{J}; \quad c = \frac{R}{L}; \quad c_1 = \frac{R_1}{L_1}; \quad d = 2\sqrt{2}\frac{nBS}{L} \\ \gamma_{max} = 2\sqrt{2}\frac{nBSe}{JR}; \quad \gamma_1 = \frac{M_1}{J} \\ \delta_1(\vartheta) = \gamma_{max}\sin\vartheta - \gamma_1 \end{gathered} \tag{21}$$

In addition, it should be noted that in the equations specified by the number (20) in its last expression the magnitude does not affect the other differential equations and the stability of the induction motor. Therefore, the following considerations only consider the following mathematical relations

$$\begin{gathered} \frac{d\vartheta}{dt} = s \\ \frac{ds}{dt} = ax\sin\vartheta + bv - \delta_1(\vartheta) \\ \frac{d\mu}{dt} = -c_1\mu + vs\sin\vartheta \\ \frac{dx}{dt} = -c_1 x + ds \\ \frac{dv}{dt} = -c_1 v - s\mu - s \end{gathered} \tag{22}$$

Subsequently, considering the change of permanent magnet coordinates, the following were obtained

$$\begin{gathered} \vartheta = \frac{\pi}{4} - \theta \\ s = -\omega \\ x = -\frac{2L}{nSB}\left[-\left(i_1 + i_3 + \frac{2e}{L}\right)\sin\theta - \left(i_2 + i_4 + \frac{2e}{L}\right)\cos\theta\right] \\ y = -\frac{2L}{nSB}\left[-\left(i_1 + i_3 + \frac{2e}{L}\right)\cos\theta - \left(i_2 + i_4 + \frac{2e}{L}\right)\sin\theta\right] \\ \mu = -\frac{2L_1}{n_1 r_0 l_1 B}\sum_{k=1}^{n_0} i_k \sin\left(\theta + \frac{2k\pi}{n_1}\right) \\ v = -\frac{2L_1}{n_1 r_0 l_1 B}\sum_{k=1}^{n_0} i_k \cos\left(\theta + \frac{2k\pi}{n_1}\right) \\ z_1 = i_1 - i_3 \\ z_2 = i_2 - i_4 \\ z_k = \sum_{k=-\frac{n_1}{4}}^{\frac{n_1}{4}} i_{(k+j)} + i_k ctg\frac{\pi}{n_1}, \quad k = 3, \dots, n_1 \end{gathered} \tag{23}$$

The equations presented (23), as in the previous considerations, were reduced to the following form

$$\begin{gathered} \frac{d\vartheta}{dt} = s \\ \frac{ds}{dt} = ay + bv - \delta_1(\vartheta) \\ \frac{d\mu}{dt} = -c_1\mu + vs \\ \frac{dx}{dt} = -c_1 x + ys \\ \frac{dv}{dt} = -c_1 v - s\mu - s \\ \frac{dy}{dt} = -c_1 v - xs - s \\ \frac{dz_k}{dt} = -c_1 z_k, \quad k = 3, \dots, n_1 \\ \frac{dz_1}{dt} = -c_1 z_1 \\ \frac{dz_2}{dt} = -c_1 z_2 \end{gathered} \tag{24}$$

where

$$\begin{gathered} a = \frac{(nBS)^2}{JL}; \quad b = \frac{n_0(S_0B)^2}{JL_0}; \quad c = \frac{R}{L}; \quad c_1 = \frac{R_1}{L_1}; \\ \gamma_{max} = 2\sqrt{2}\frac{nBSe}{JR}; \quad \gamma_1 = \frac{M_1}{J} \\ \delta_1(\vartheta) = \gamma_{max}\sin\vartheta - \gamma_1 \end{gathered} \tag{25}$$

In addition, it should be noted that the variable z_k has no effect on the stability of the synchronous machine, so the differential equations describing engine operation can be written as follows

$$\begin{gathered} \frac{d\vartheta}{dt} = s \\ \frac{ds}{dt} = ay + bv - \delta_1(\vartheta) \\ \frac{d\mu}{dt} = -c_1\mu + vs \\ \frac{dx}{dt} = -c_1 x + ys \\ \frac{dv}{dt} = -c_1 v - s\mu - s \\ \frac{dy}{dt} = -c_1 v - xs - s \end{gathered} \tag{26}$$

Thus, the mathematical equations describing the dynamics of a 4-pole motor with a damping winding at different types of connection are limited only to the mathematical relations represented by the expressions (22) and (26).

3 Simulation of selected components of the ASE system according to the concept of MEA

Derived mathematical relationships (model) describing the physical phenomena occurring in the PMSM synchronous electric motor (engine) were subjected to a simulation verification performed in the Matlab/ Simulink programming environment [14], [15]. In the next step, using the available mathematical functions and blocks to map the individual electronic components, a synchronous motor model was developed for the operation of the electrical network of the aircraft [16]. The basic technical parameters characterizing the synchronous electric machine PMSM in the form of an electric motor were selected according to units referenced to the three-phase permanent magnet synchronous motor rating, namely $R_s = 0.2\Omega$; $L_d = L_q = L_s = 0.0085H$; $\Psi_{PM} = 0.175$ Vs; p = 4. The magnitude of R_s the substitute resistance of the motor circuit; and L_d, L_q, L_s are the inductances of the coils contained in the motor. In turn, the parameter Ψ_{PM} is responsible for the value of the electromagnetic flux, and p is the number of permanent magnets. Synchronous motor block diagram designed in Matlab/Simulik is shown in Fig. 6.

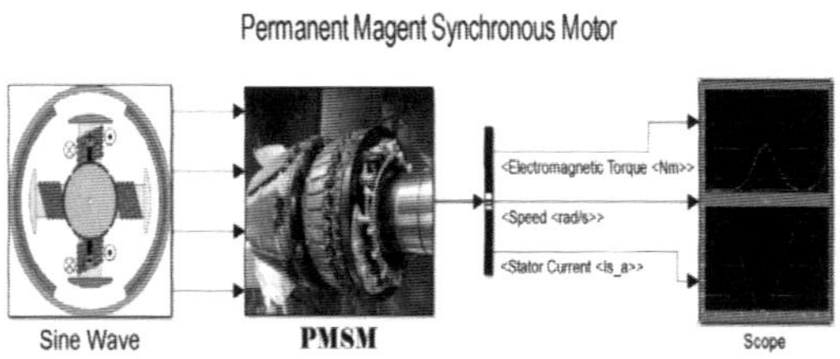

Fig. 6. Model of synchronous motor in Matlab/Simulink program

It is important to note that during the start-up of a permanent magnet-driven motor, the position of the stator field and rotor field axes is important. This can also be demonstrated by analyzing the rotational speed during start-up of the motor for the various phases of the stator start-up phase. The next figure (Fig. 7) shows the engine rotational speed charts for the extreme rotor positions relative to the initial supply voltage phase of 0 and 180°. The results of the simulations are shown in Figs. 7 to 9.

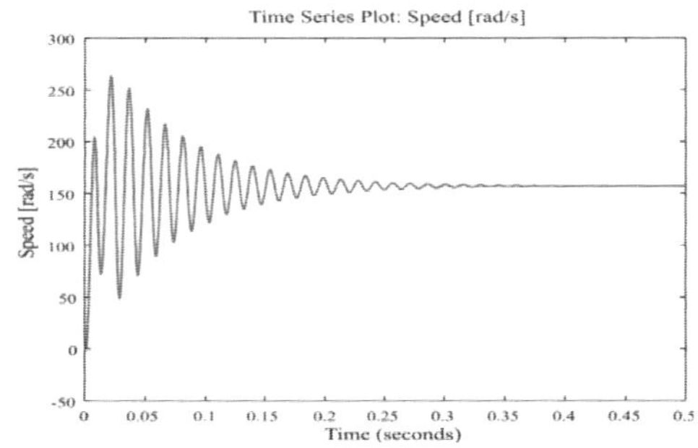

Fig. 7. Rotational speed diagram of the synchronous motor

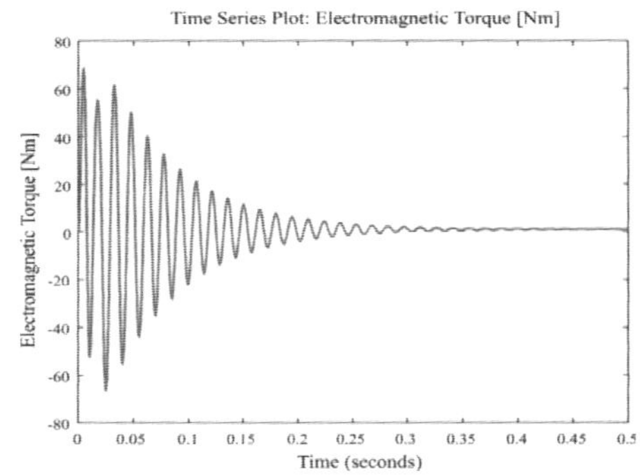

Fig. 8. Electromagnetic torque diagram of synchronous motor

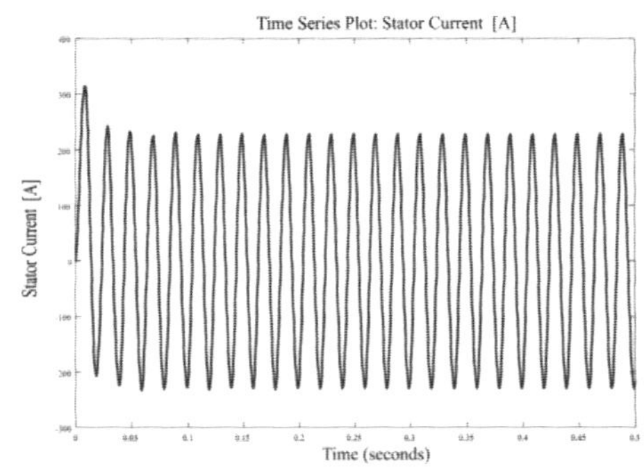

Fig. 9. Diagram of current waveform in synchronous motor

Presented waveforms representing the electromagnetic torque and rotational speed occurring during the start-up process and during further operation of the motor acquire the characteristic waveform of a sinusoidal wave with varying amplitude and duration, but with a constant frequency value. The shape of the curves in both cases resembles an oscillating signal which, after a certain time, converts into a straight line. In addition, the current

characteristic of the synchronous motor circuit is a curve that is characterized by some fluctuations at the very start of the motor start-up, followed by a sinusoidal waveform. It should be added that the resulting initial fluctuations result from the starting of an electric motor. The start-up properties of PMSM electrical machines have a direct impact on their operation. In turn, the braking torque from permanent magnets, occurring in the lower speed range, significantly impedes the start-up process. Thanks to the measured short-circuit characteristics, the elements of the calculation model (such as face-inductance and material parameters) were significantly modified from the standpoint of start-up characteristics, followed by average torque values and static mechanical characteristics (Fig. 8).
Additionally, the graph shows that the starting torque of the motor is sufficiently large, even when the voltage is applied, including the possible voltage drop (17%). Calculations were made assuming that the engine was loaded with electronic equipment (load compatible with the technical requirements according to the standards used in more electric aircraft) for the moment of inertia equal to 1.2 the moment of inertia of the motor rotor of the PMSM electrical machine.

4 Conclusions

Modern synchronous motors with permanent magnets in the electrical network of aircraft compatible with the concept of a more electric aircraft have very good regulatory characteristics and are characterized by high efficiency and high torque overload. These properties are confirmed by the diagrams shown in Figs. 7 to 9. In addition, it should be added that high torque momentum becomes a key parameter for the selection of motors in electrical drives that produce electricity.
Because of the motor nature of the high instantaneous overload, the mass and rated capacities of the motors used in electric aviation engines can be reduced to a large extent. In addition, the use in permanent magnet motors of standard spread sheets used in asynchronous motors can significantly reduce the cost of production of these motors, but they are not optimized for maximum torque. Consequently, stator and rotor plates must be designed to achieve the best operating parameters of permanent magnet motor. However, it is not feasible to design such a motor that would have all the key technical parameters at a very high level. In addition, motors for various applications will have different parameters and the designer should decide what parameters are the most important for an engine to adequately design the motor electromagnetic circuit (Figs. 7 to 9).
Current permanent magnet engine designs are strictly dedicated to a particular drive, where the most important parameter is the engine mass or power density from the unit of volume. To make full use of the power of permanent magnet motors, the motor should be powered by a dedicated power converter. Based on the obtained simulations it can be stated that by designing a synchronous motor, its properties can be independent of possible changes of the mains voltage (this includes e.g. power factor). This phenomenon plays a special role in power supply networks in aircraft in line with the trend of more electric aircraft [17]. In addition, be sure to include power supply impedance and voltage drop during design to ensure (guarantee) sure start-up and synchronization of the motor. Furthermore, since the design of such an engine, adapted to specific power and load conditions, is tailored to individual requirements, it is characterized by high energy efficiency with maximum efficiency and power factor.

References

1. Abdel-Fadil R., Eid A., Abdel-Salam M., Electrical distribution power systems of modern civil aircrafts, 2nd International Conference on Energy Systems and Technologies, Cairo, Egypt 2013.
2. Setlak L., and Kowalik R., Mathematical modeling and simulation of selected components on-board autonomous power supply system (ASE), in accordance with the concept of a more electric aircraft (MEA), 2017 18th International Scientific Conference on Electric Power Engineering (EPE) Kouty nad Desnou, Czech Republic, pp. 1-6, IEEE 2017.
3. Moir I., Seabridge A., Design and Development of Aircraft Systems, Second Edition, John Wiley & Sons, Ltd., 2013.
4. Krause P.C., Wasynczuk O., Sudhoff S.D., Analysis of Electrical Machinery and Drive Systems, IEEE Press, John Wiley & Sons, Inc., 2002.
5. Parker R., Horizon 2020 Industrial Technologies as an Engine of Growth, Rolls-Royce, Presentation 2012.
6. Zhao X., Guererro J.M., Wu X., Review of Aircraft Electric Power Systems and Architectures, International Energy Conference (ENERGYCON), IEEE 2014.
7. Setlak L., Overview of Aircraft Technology solutions compatible with the concept of MEA, Technical Transactions. Electrical Engineering, No. 1-E/2015, p. 67-76, Cracow University of Technology 2015.
8. Toliyat H.A., and Kliman G.B., Handbook of Electric Motors, Taylor and Francis. Electrical and computer engineering, 850, (2010).
9. Singh B., Murthy S.S, and Gupta S., Analysis and design of STATCOM-based voltage regulator for self-excited induction generators, IEEE Trans. Energy Convers., Vol. 19, No. 4, pp. 783–790, (2004).
10. Wang L., Jatskevich J., and Dommel H.W., Re-examination of synchronous machine modeling techniques for electromagnetic transient simulations. Power Systems, IEEE Transactions on 22 (3), 1221-1230, (2007).
11. Yang Z., Qu J., Ma J., and Shi X., Modeling and Simulation of Power Distribution System in More Electric Aircraft, Hindawi Publishing Corporation

Journal of Electrical and Computer Engineering Volume, Article ID 847624, 7 pages, 2015.
12. Cheng R., Zhao W., Deng H., and Jiang X., Modeling and Optimization Control for Aircraft AC Generator Brushless Excitation System Based on Improved Adaptive PSO, The Open Automation and Control Systems Journal, Vol. 7, 21-30, (2015).
13. Tesla N., No. 382,280, Electrical Transmision of Power, 1888.
14. Vinod Vashram Vadher, Mathematical Modelling of Aircraft Electrical Power Systems, Doctoral Thesis, Loughborough University of Technology, 1981.
15. Skvarenina T.L., Pekarek S., Wasynczuk O., and Krause P.C., Simulation of a More-Electric Aircraft Power System using an automated state model approach, Proceedings of the Intersociety Energy Conversion Engineering Conference, pp. 133-136, (1996).
16. Emandi K., Ehsani M., Aircraft power systems: technology, state of the art, and future trends, Aerospace and Electronic Systems, IEEE, 2000.
17. Setlak L., Kowalik R., Comparative Analysis and Simulation of selected Components of Modern On-board Autonomous Power Systems (ASE) of Modern Aircraft in line with the Concept of MEA/AEA, Lecture Notes in Engineering and Computer Science, Volume 1, 2016.

ITM Web of Conferences **16**, 03002 (2018) https://doi.org/10.1051/itmconf/20181603002
AMCSE 2017

Model Residual Stress by Finite Element Method

Delia Garleanu[1,*], *Claudia* Borda[1], *Gabriel* Garleanu[1], and *Victor* Popovici[1]

[1]UPB, IMST Faculty, TMS Department, 060042 Splaiul Independentei 313, Bucharest, Romania.

Abstract. This paper presents an original model developed by finite element method to simulate the behavior of the material to the method "Blind Hole Drilling", to determine the residual stress. Modeling of this method is possible through the use of the "Birth and Death" which have some elements of ANSYS library. After obtaining the analysis of movements, appropriate loads, a node located from the center hole at a radius calculated. In this way it is easier to estimate the stresses and deformations of a piece. Several measurements are made and based on this model is given in ANSYS. In this way we can have a map of tensions and deformations in a material

1 Introduction

One of the most common methods of residual stress measurement technique is the method " Blind Hole Drilling " .

Underthis method, after installingthepieceof sensors (strain gauges) is practicedshallowholeand surfacemodificationstate ofdeformationmeasuredin the vicinity ofthe hole.Fromexperimental dataobtainedcan be calculatedresidual stressesrelaxed(released).

This method is consideredsemidestructiveas long as thesmallholeinmost casesnotsignificantly affectthe structural integrityof the partbeing tested.It is atypicalholeD$_0$ =1/16"=1,5mmandD$_0$ = 1/8" = 3,0mmboth indepthanddiameter.If necessary, the hole canbe coveredafter measurement. This is astandardtest method, developedby the AmericanSociety forTestingand Materials(ASTM 837). Figure 1 shows the drilling device RS-200 model.

Fig. 1. End-Mill Drilling RS-200 model.

2 Principle of the method "Blind Hole Drilling"

Practice of hole (even very small diameter) in a body residual stress, relax tension in that place because the normal stress at any free surface (in this case the surface hole) must necessarily be zero. Elimination normal stress at the edge of the hole reduce tensions in the region surrounding vicinity, causing local deformations to be amended accordingly. Measuring strain changes caused by stress relaxation provides the necessary data for calculation. When you take avery small hole diameter (fig.2) in a region containing residual stresses, deformations sizes at point P are functions of the main local stress σ_p and σ_q and the geometric relationship between point and hole between point and principal axes.

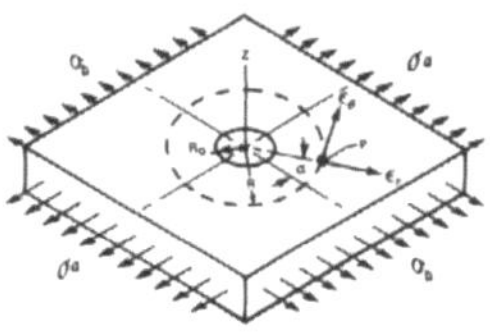

Fig. 2. Strain relaxation at point "P" due to a hole practice.

For example, if one considers deformations of the point where the main stressis present only σ_p, then we can write the relations:

$$\varepsilon_r = -\sigma_p \frac{1+\nu}{2E}\left[\frac{1}{r^2} - \frac{3}{r^4}\cos 2\alpha + \left(\frac{4}{1+\nu}\right)\frac{1}{r^2}\cos 2\alpha\right] \quad (1)$$

$$\varepsilon_\theta = -\sigma_p \frac{1+\nu}{2E}\left[-\frac{1}{r^2} + \frac{3}{r^4}\cos 2\alpha - \left(\frac{4\nu}{1+\nu}\right)\frac{1}{r^2}\cos 2\alpha\right] \quad (2)$$

* Corresponding author: delia.garleanu@upb.ro

$$\gamma_{r\theta} = \frac{\sigma_p}{2G}\left(\frac{3}{r^4} - \frac{2}{r^2}\right)\sin 2\alpha \quad (3)$$

$$\varepsilon_r - \varepsilon_\theta = -\sigma_p \frac{1+\nu}{2E}\left(\frac{2}{r^2} - \frac{6}{r^4}\cos 2\alpha + \frac{4}{r^2}\cos 2\alpha\right) \quad (4)$$

where: ε_r, ε_θ are tangential or radial deformations at point P; $\gamma_{r\theta}$ - shear strain at point P, $r=R/R_0$ dimensionless distance from the hole center point P, E, G, ν - module of elasticity, transverse and Poison coefficient of the material. Deformations revealed relationships expressed above are shown in figure3 for $\alpha = 0^0$ and $\gamma = 90^0$ to illustrate their variation on the main axes directions in relation to the distance (dimensionless) from the center hole.

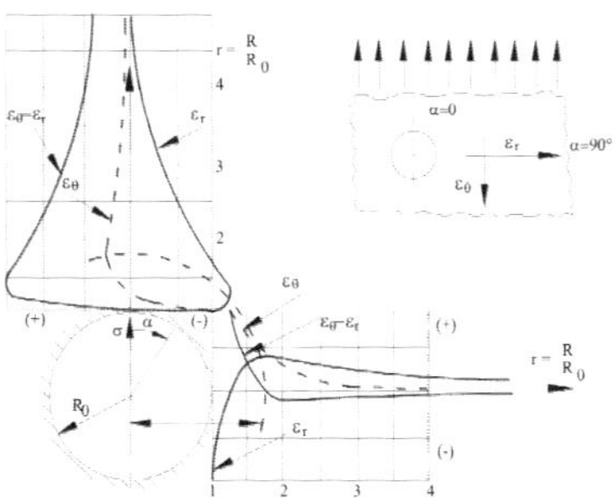

Fig. 3. Changes highlighted strains distance from the hole center (along-axis) -uniaxial residual stress.

As the distance from the hole increases, deformations decline rapidly. For this reason, to obtain a maximum signal strain gauges, it is preferable that the measurement be made as close to the hole.

On the other hand, increasing parasitic effects in the immediate vicinity of the hole. These considerations lead to the understanding of compromise in selecting the optimal range.

As shown in figure 4 is installed three rosette strain gauges oriented radially to the center of the radius R from the center hole. We recommend the use of specialized products rosettes "Micro-Measurements Division". Rosettes are available in two versions (R =0.101in, R =0.202in) with different types of temperature compensation.

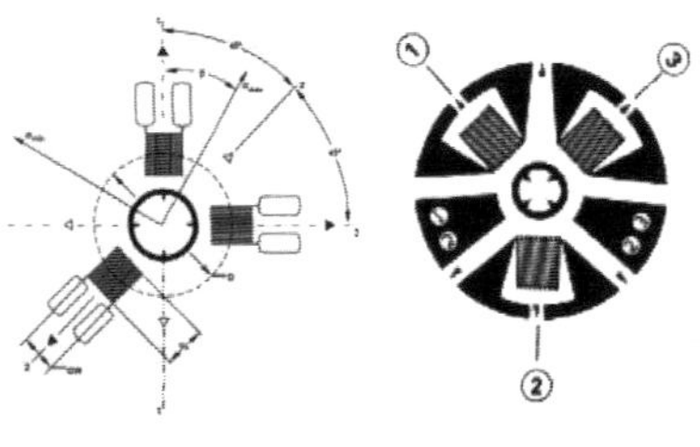

Fig. 4. Positioning rosette strain gauges for determining residual stress.

3 Modeling by finite element method "Blind Hole Drilling"

Finite element modeling method "Blind HoleDrilling" for determining residual stress a material can be done using the feature "Birth and Death" (activation-deactivation), which have some elements of ANSYS meshingthe library.

The "Birth and Death" every element meshis used to activate and deactivate them when material is added (or removed) from the system. Thus, some elements of the modeling "existing" or "absent". To achieve the "element death" ANSYS program does not remove disabled items but to reduce stiffness. Although the loads associated disabled elements are brought to zero, they appear in the list of elemental charges.

3.1. Creating model

Design geometry model begins by building a parallelepiped with dimensions of 10x10x4mm. With in this volume are defined six volumes rectangular cylinder diameter and total depth of 3mm(fig. 5a).

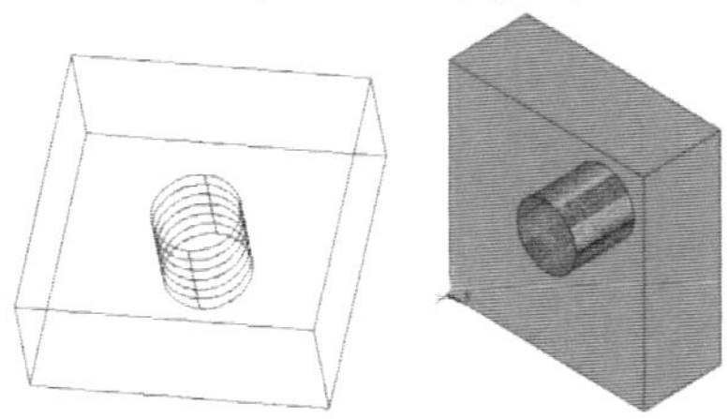

Fig.5.a Creating volumes.

Subsequently, these six volumes elements are assigned properties "Birth and Death" so that it can be modeled and incremental "Blind Hole Drilling" method. Inside this volume are defined rectangular cylindrical diameter and six volumes total depth of 3mm (Fig.5 b).

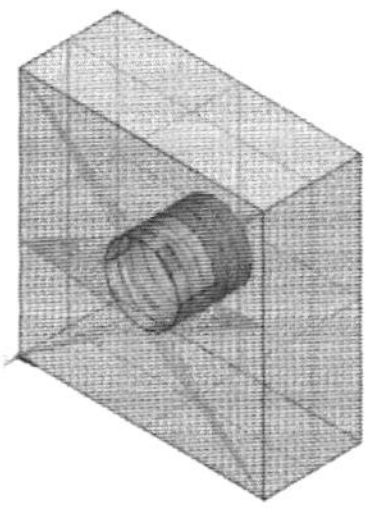

Fig.5b. Six cylindrical volume to be removed successively-incremental method

Meshing volumes (fig.6) using SOLID92 element determines obtain items totaling 16,989 nodes.

 https://doi.org/10.1051/itmconf/20181603002

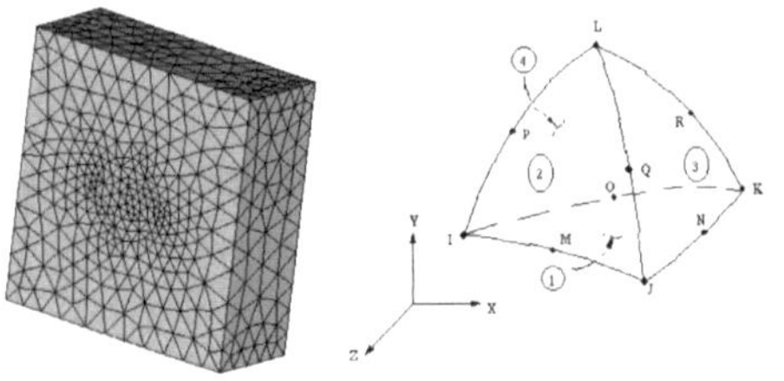

Fig. 6. Mesh elements parallelepiped volumes (SOLID 92).

In table 1 is indicated the number of cylindrical volumes, the depth and the relevants elements.

Table 1. Hole depth and volume elements.

No.volume	Depth [mm]	Elements
1	0,5	1-531
2	0,1	532-1107
3	1,5	1108-1676
4	2,0	1677-2245
5	2,5	2246-2815
6	3,0	2816-3389

3.2 Application loading

Choose the type of static analysis. For induction into the fabric of a state of biaxial tension flat areas apply lateral load displacement type (fig.7) with values between 0.01-0.1mm in six steps(0.01, 0.025, 0.05, 0.075, 0.1mm).

The solution runs through seven successive runs. The first run does not disable any corresponding element cylindrical volumes. For each of the six following analysis is off by a volume cylindrical elements.

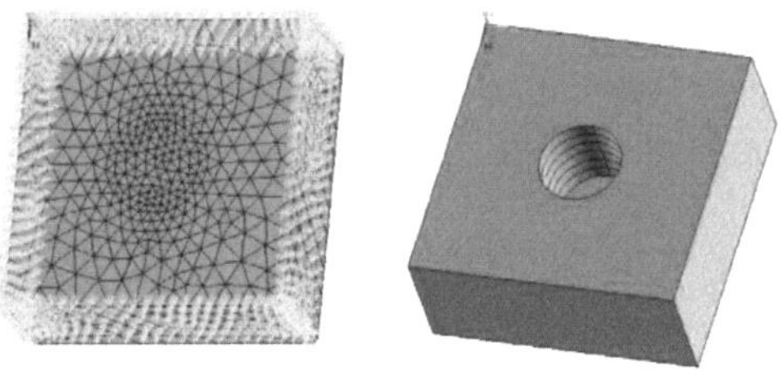

Fig.7. Application loading and off elements.

In figure 8 are represented three of the six off successive phases corresponding cylindrical hole data volumes incremental material.

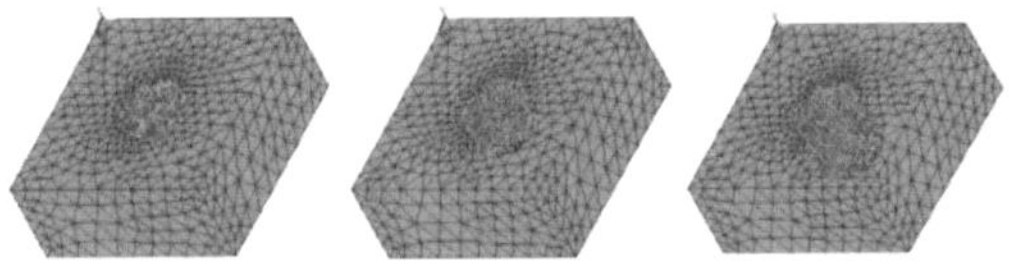

Fig. 8. Disabling successive elements of the six volumes of hole.

Finite element modeling method "Blind HoleDrilling", to determine the residual stressin a material seeking a node displacements values found on the material at a radius calculated according to the theory of strength of materials by R_{ph}= 2,598 mm from the center hole. For this purpose (fig.9), select the area on the surface of box volume (Area 41), which is measuring the deformations that occur increasing tensions by playing the hole (in a single pass at a depth equal to the diameter of the hole ori ncremental).In this area, choose a node (Node 6192) located at the radius corresponding to the calculated (2.598mm). The tables with the results calculated by the processor selects the direction OX nodal displacement values corresponding to this node.

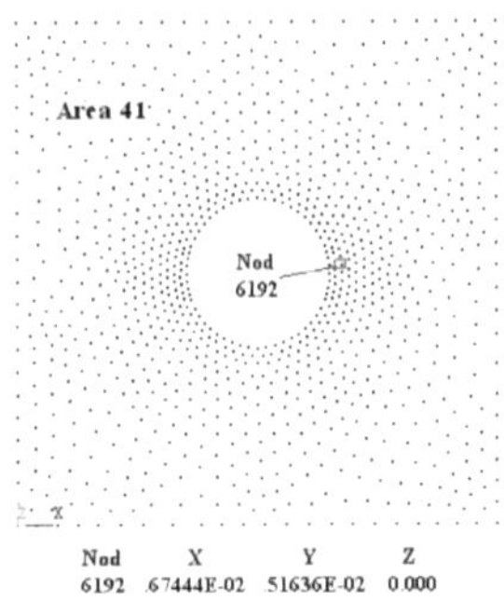

Fig.9. Selecting Node 6192 on Area 41.

The results displacements for node 6192 adequate to loads applied displacement volume of rectangular side areas that will induce tension are given in table 2.

Table 2. Nodal displacement values.

Displacement load (±UX)/(±UY) [mm]	Displacement UX nod.6192 [mm]
1 E-02	0.0076839
2.5 E-02	0.0019258
5 E-02	0.0038677
7.5 E-02	0.0058258
10 E-02	0.0078001

Graphical representation of the load-displacement pairs of values shows a very good linearity. Using a spreadsheet program is determined linear equation (fig10).

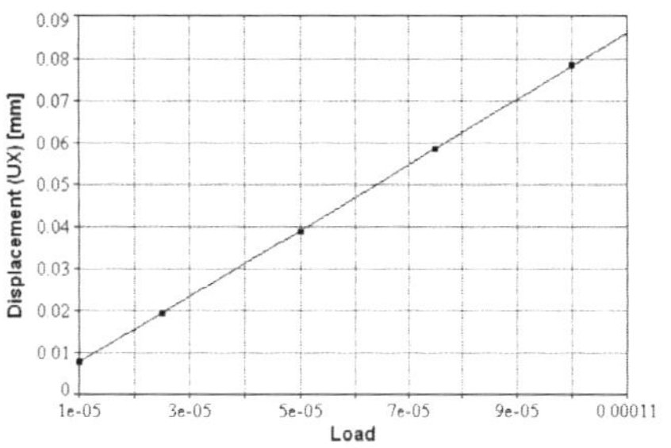

Fig. 10. Dependency relationship load - displacement.

Results strains (vonMises and Intensity) for node 6192, in order deactivation of successive volumes of the six corresponding hole until it reaches a depth value equal to its diameter are given in table 3. Figure11 is plotted the variation deformation (vonMises and Intensity) for node 6192, using the data from table3.

Table 3. Values corresponding deformations volumes off

Deactivated volume	Node no.	Von Mises strain	Intensity strain
0	0	0.009402	0.00942
1	1-531	0.011446	0.011604
2	532-1107	0.013074	0.013863
3	1108-1676	0.014003	0.015041
4	1677-2245	0.014507	0.015504
5	2246-2815	0.014623	0.015564
6	2816-3389	0.014537	0.015554

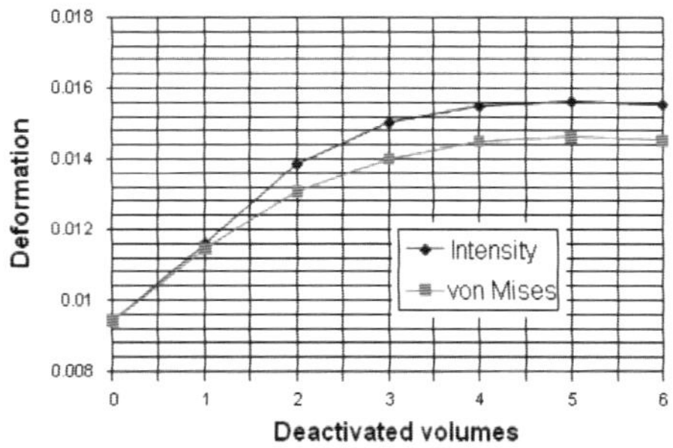

Fig. 11. Variation of deformation (von Mises and Intensity) with volume off.

In figure 12 are plotted nodal results, von Mises-section deformations resulting from successive runs that was disabled by a volume of the hole, thus simulating method "Blind Hole Drilling" incremental.

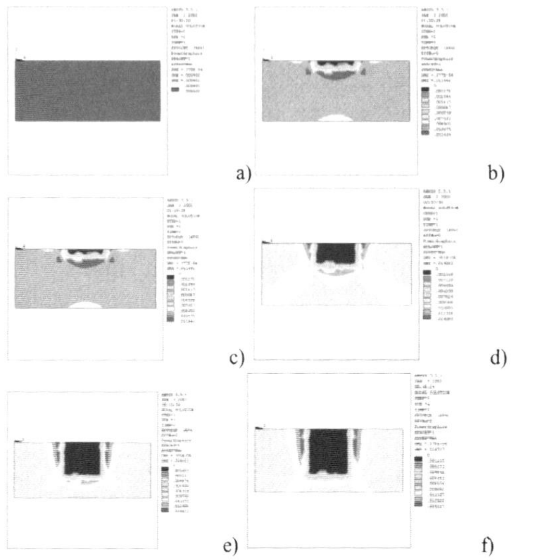

Fig.12. Representation of nodal results, vonMises strain-section for off successive volumes of the cylindrical hole.

To confront the results obtained from finite element modeling "Blind Hole Drilling" method with values derived from the methodology of calculation was calculated the percentage ratio of strain to the ratio depth/diameter (Z/D) equal to off successive volumes the cylindrical hole. Results are shown in table 4.

Table 3.

Deactivated volume	Z/D	Von Mises strain	Percentage ratio of strain	
			Strain difference	%
0	0	0.009402	0	0
1	0.166	0.011446	0.002044	39.149
2	0.333	0.013074	0.003672	70.331
3	0.5	0.014003	0.004601	88.125
4	0.666	0.014507	0.005105	97.778
5	0.833	0.014623	0.005221	100
6	1.0	0.014537	0.005135	98.353

Graphical representation of the values in Table 5 is given in Figure 13.

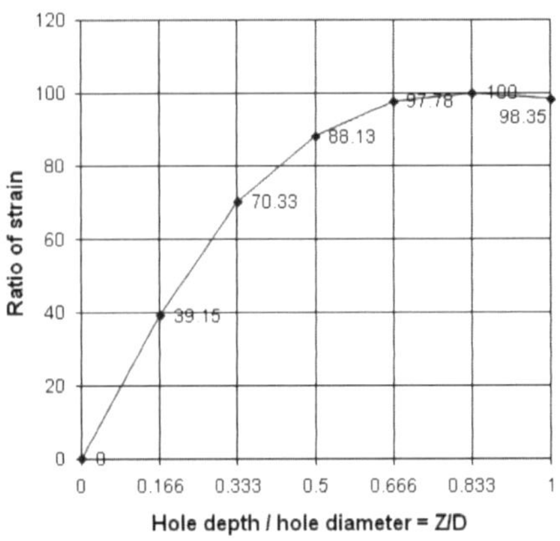

Fig.13. Ratio of strain to the ratio Z/D (depth/diameter)

4 Conclusion

Presented an original model developed by finite element method that simulates the behavior of the material to the method "Blind HoleDrilling", to determine the residual stress. Modeling of this method is possible through the use of the "Birth and Death" (activation-deactivation), which have some elements of ANSYS meshing library. To get the effect "element death", elements are cutoff rigidity. After the analysis results are obtained displacements corresponding loads, a node located from the center hole at a radius calculated. To confront the results of modeling by Finite Elements Method, method "Blind Hole Drilling" with values from the calculation methodologyis calculated "the percentage strain ratio" the ratio depth / diameter, resulting in deactivation of successive volumes of the cylindrical hole. In industry, the percentage of use of welded constructions / welded parts is very high. After each welding cord we have strains and deformations. Deformations are easy to see because they are visible and measurable. The state of stress in the material is determined more difficult and depending on the importance of the part, the thermal stress relief treatment is done or not. Thermal treatment involves some costs. It is necessary to know what is the level of tension in the material in the whole piece to make the decision to achieve or not the heat treatment of the piece. It is difficult to make hundreds of measurements on the same piece and therefore a finite element program can be used to simulate the residual stresses in the piece.

References

1. Measurement of Residual Stresses by the Hole-Drilling Strain Gage Method.-Tech Note TN 503-6, Vishay Micro-Measurements
2. Standard Test Method for Determining Residual Stresses by the Hole-DrillingStrain-Gage Method." ASTM Standard E 837-01 2001
3. G. Gârleanu, Computerization and optimization of technological processes, Bucuresti, 2007.
4. D. Gârleanu, "Pressure Welding Technology," Ed.Bren, Bucuresti 2014
5. W.H. Lee, *Computer Simulation of Shaped Charge Problem* Ed. World Scientific Publishing Co. Pte. Ltd.,2006.
6. P., Krusl, *A Pragmatic Introduction to the Finite Element Method for Thermal and Stress Analysis*, Ed. World Scientific Publishing Co. Pte. Ltd., 2006
7. ANSYS Tutorials
8. Witt, F., F. Lee and W. Rider "*A Comparison of Residual Stress Measurements Using Blind-hole Drill, Abrasive Jet, Trepan Ring*." Presented at Society for Experimental Stress Analysis Meeting, Dearborn, Michigan, 1981
9. Measurement Group, Inc."*Measurement of Residual Stress By The Blind Hole Drilling Method*" 2001
10. Rendler, N.J. and I.Vigness. "*Hole-drilling Strain-gage Method of Measuring Residual Stresses*". Proceedings of the Society of Experimental Stress Analysis. Vol. XXIII, No.2, 1996, pp. 577-586

 https://doi.org/10.1051/itmconf/20181603003

Calculation and Experimental Validation of Pressure and Temperature Effects on COG-Air Fuel Mixtures

Skrinsky Jan

Energy Research Centre, VŠB-TU Ostrava, 17. Listopadu 15/2172, 708 33 Ostrava, Czech Republic

Abstract. COG have been widely used together with blast furnace gas and blast furnace oxygen gas in the steel industry in Moravian-Silesian region of Czech Republic. COG is a flammable and explosive substance. Most explosion characteristics published so far are valid for pure compounds and limited experimental conditions, mostly ambient. There have been no explosion characteristic exists for COG-air mixtures which cover industrial conditions up to 423 K. Experimental tests have been carried out in a 20-L closed explosion chamber adopted for the explosion tests. The element potential approach in the thermochemical equilibrium calculations applied in the Chemkin subroutine has been used for explosion pressure calculations. Different explosion characteristics have been reported in a range from 298 K up to 423 K and from 0.5 bar(a) up to 1.0 bar(a).

1 Introduction

All process industry accidents fall under three broad categories fire, explosion, and toxic release. Of these, fire is the most common, followed by explosions. Within these categories explosion make the immediate damage potential [1].

Coke oven gas (COG) is highly rated as a valuable by-product of coal carbonization to produce coke in the steel industry [2].

Although COG is regarded as a non-standard gaseous fuel, it still has a reasonable energy content and calorific value, which depend on the nature of coal and the type of carbonization and have been widely used together with blast furnace gas and converter gas in the steel industry in Moravian-Silesian region of Czech Republic. However, coke oven gas is a flammable and explosive substance. Improper operation in the process of production, usage, transportation and storage can easily lead to combustion and explosion, posing a serious threat to lives and property. Therefore, how to safely and effectively utilize or dispose of coke oven gas has become an urgent task [3].

There are no explosion characteristic exists for COG which cover all conditions occurring in "real life" initial conditions. Most explosion characteristics published so far are valid for specific and limited conditions. Few data at standard conditions are available in open literature [4]. In the presented paper these characteristics have been systematically examined and determined in the 20-L explosion chamber for the first time.

2 Experiment

2.1 Experimental device

Measurements of the P_{max} and $(dP/dT)_{max}$ have been performed in the 20-L explosion chamber (OZM Research s.r.o., Czech Republic). The spherical explosion chamber allows the measurement of parameters of explosive gas dispersion in accordance with the terms and specifications in EN 15967 [5]. The mixing process proceeds inside the explosion chamber. The gasses are dosed into the chamber from the dosing vessel equipped by three inputs for connection to cylinders with nitrogen and oxygen and line for free suction of air from outside. The ignition source is located in the geometrical centre of the cylinder. In the experiment, ignition was performed with hot wire between stainless steel electrodes.

2.2 Experimental procedure

The experimental procedure was as follows: (a) the air in the explosion vessel was evacuated to a vacuum of 0.1 mbar; (b) mixing gases were added into the vessel at different ratios; (c) the mixture was admitted at the desired pressure, then ignited at once; (d) after ignition and the capture of the signals by the acquisition system, the burned gas was completely evacuated. Then, a new cycle was repeated.

* Corresponding author: author@e-mail.org

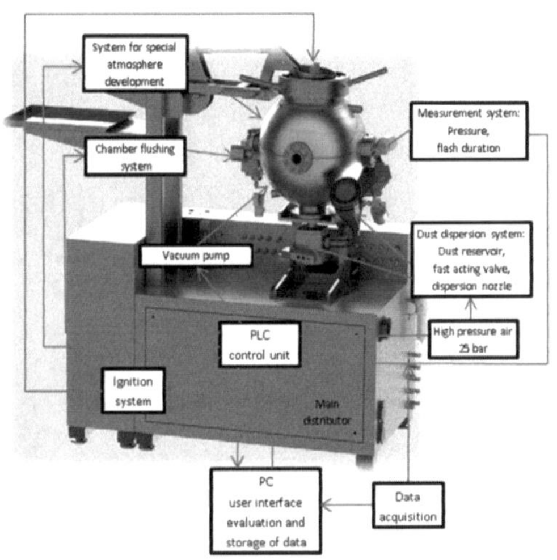

Fig. 1. Spherical vessel experimental arrangement.

3 Calculation

Calculation procedure is described in [3]. The element potential approach in the thermochemical equilibrium calculations applied in the Chemkin subroutine has been used for explosion pressure calculations. This approach represents 'ideal' deflagrations in closed systems well and gives the highest possible attainable explosion pressures. Explosion pressures are calculated using the species and their thermodynamic values from the GRI 3.0.

The maximum pressure rise rate during gas explosions in enclosures, $(dP/dt)_{max}$, and the deflagration index, K_G, are important explosion characteristics of premixture. They are used to quantify the potential severity of an explosion.

The maximum pressure rise rate, $(dP/dt)_{max}$, depends not only on the mixture properties (such as mixture composition, initial temperature and initial pressure) but also on the volume of the vessel in which gas explosion takes place. Unlike (dP/dt)max, the deflagration index is an intrinsic property of the premixture and it is independent of the volume of the vessel used in experimental measurements.

The relationship between K_G and $(dP/dt)_{max}$ is given by Eq(1):

$$K_G = (dP/dt)_{max}\, V^{1/3} \qquad (1)$$

in which K_G is the deflagration index (bar.m/s), V is the vessel volume (m^3), dP/dT is the maximum rate of pressure rise (bar/s). In order to capture the intrinsic reactivity of COG mixture, the burning velocity has been obtained from time pressure records of explosions occurring in 20-L closed explosion chamber.

The burning velocity in Eq(3) first published by [6] has been calculated from the pressure time history by using the time derivative of flame radius, r_f, as given by the correlation in Eq(2):

$$r_f=(3V/4\pi)^{1/3}\,[1-(P_0/P)^{1/\gamma}\,((P_{max}-P)/(P_{max}-P_0))]^{1/3} \qquad (2)$$

in which r_f is the flame radius (m), V is the vessel volume (m^3), P_0 is the initial pressure (bar), P is the actual pressure (bar), P_{max} is the maximum explosion pressure (bar) and γ the adiabatic coefficient of the unburned gas (-).

$$s = [1/((P_{max}-P_0))^{1/3}(4\pi/3V)^{-1/3}\,(P/P_0)^{1/\gamma}\,[1-(P/P_0)^{1/\gamma}\,((P_{max}-P)/(P_{max}-P_0))]^{-2/3}\,dP/dt \qquad (3)$$

in which s is burning velocity (m/s), r_f is the flame radius (m), R is specific gas constant (J/kg.K), V is the vessel volume (m_3), P_0 is the initial pressure (bar), P is the actual pressure (bar), P_{max} is the maximum explosion pressure (bar), γ the adiabatic coefficient of the unburned gas (-) and dP/dT is the rate of pressure rise (bar/s).

4 Results and discussion

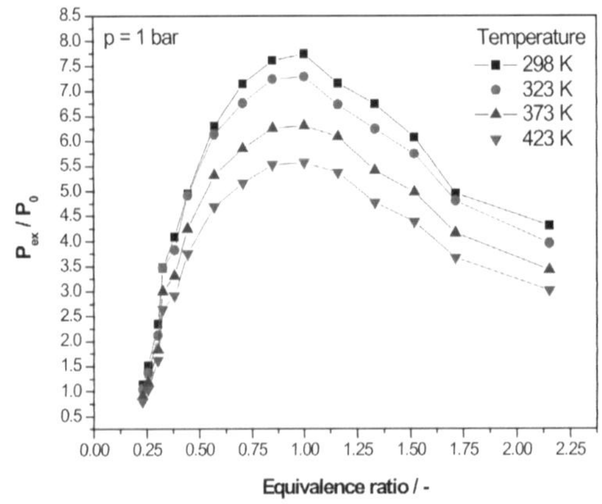

Fig. 2. Normalized peak explosion pressure (P_{max}/P_0) versus equivalence ratio at atmospheric pressure and four initial temperatures.

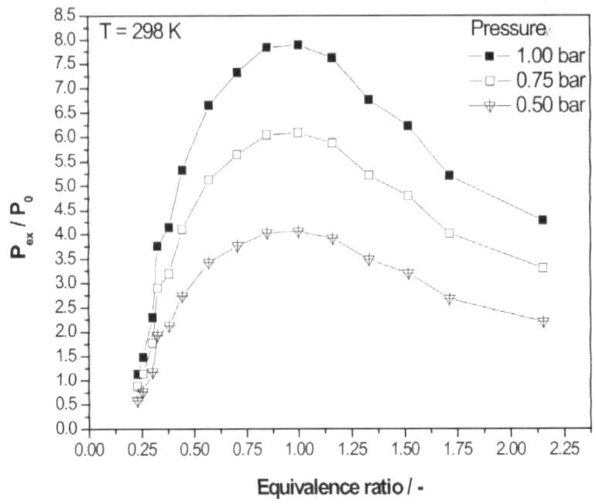

Fig. 3. Normalized peak explosion pressure (P_{max}/P_0) versus equivalence ratio at atmospheric temperature and three initial pressures.

As observed from Fig. 2-3 the experimental results, explosion pressure shows a certain regularity with the variation of coke oven gas concentration, and the maximum pressure achieves its peak value near the stoichiometric concentration (The calculated value based on the compositions of coke oven gas is 20.9%.) and tends to decrease if the concentration is lower or higher than 20.9%. This pattern forms because in the vicinity of the stoichiometric concentration, the coke oven gas and oxygen can be fully utilised, causing the most intense reaction to occur, and thus, the largest pressure value is generated. When the concentration of coke oven gas is less than the stoichiometric concentration, even though there is surplus oxygen, the coke oven gas is also in relatively shorter supply, so the explosion is relatively weaker, correspondingly producing relatively less pressure. When concentration of coke oven gas is more than the stoichiometric concentration, the oxygen concentration will relatively low. Thus, the coke oven gas concentration actually involved in the reaction is lower, so less pressure is produced. The greater the concentration of coke oven gas, the less oxygen content, and therefore the less of the coke oven gas actually participates in the reaction. Thus, the pressure becomes smaller.

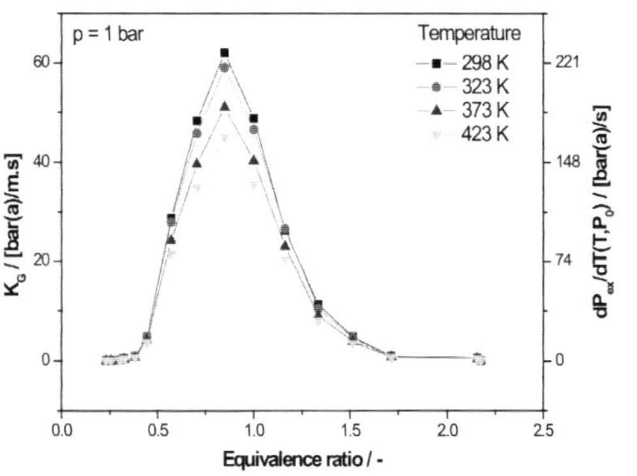

Fig. 4. Rate of pressure rise, $(dP/dt)_{ex}$, and deflagration index, K_G, versus equivalence ratio at elevated temperatures and pressures.

Fig. 4 shows the comparison among rate of pressure rise, $(dP/dt)_{ex}$, and deflagration index, K_G, versus equivalence ratio at elevated temperatures and pressures for the methanol/air mixtures. Maximum rate of pressure rise and deflagration index reach their peaks at Φ around 1.46, and they decrease at both lean and rich mixtures. As shown in Fig. 4, $(dP/dt)_{max}$ and K_G give an approximate value at varied initial temperatures, indicating that $(dP/dt)_{max}$ and K_G are sensitive to the variation of temperature (109, 102, 98, 90 bara/m.s). Rate of pressure rise is affected by both flame speed and heat release. Flame speed increases monotonically with the increase of temperature, but the decrease of total fuel mass results in the decreased heat release and offsets the effect of flame speed on the explosion pressure. With the increase of initial pressure $(dP/dt)_{max}$ and K_G increase dramatically, especially around equivalence ratio of 1.1. This differs from the flame speed, but agrees with the mass burning flux when taking into account of the effect of density. It is noted that deflagration index is less than 20 MPams^{-1} at Φ=1.0 and the initial pressure of 0.1 MPa, belonging to the first class of deflagration index and low potential of explosion hazard! However, KG could exceeds 30 MPa m s-1 around the stoichiometric ratio at the pressures higher than 0.1 MPa and could enter the highest class of deflagration index.

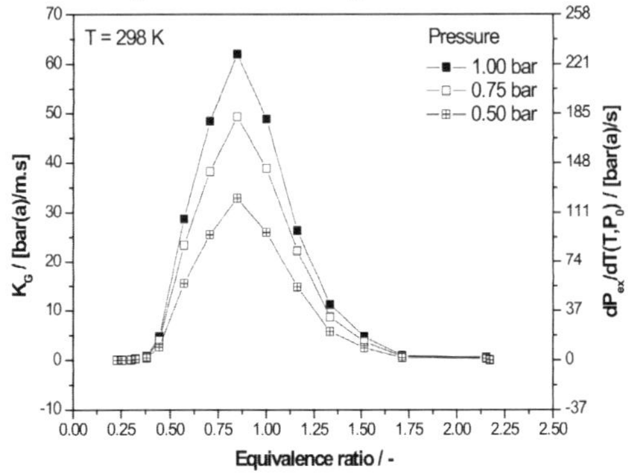

Fig. 5. Rate of pressure rise, $(dP/dt)_{ex}$, and deflagration index, K_G, versus equivalence ratio at elevated temperatures and pressures.

The maximum rate of pressure rise and the maximum explosion pressure have similar relationships with coke oven gas concentration (as plotted in Fig. 5.). When the concentration of coke oven gas is lower than its stoichiometric concentration, oxygen is relatively abundant. With the increase of coke oven gas concentration, its reaction volume per unit time increases, so the maximum rate of pressure rise also increases, and it achieves the maximum value near the stoichiometric concentration (Φ=0.9). However, when the coke oven gas concentration is higher than the stoichiometric concentration, increasing the coke oven gas concentration requires the oxygen content in the mixture to decline, so the amount of coke oven gas actually involved in the reaction is reduced correspondingly.

5 Conclusion

An experimental study on the characteristics of COG-air explosions in a 20-L closed spherical vessel with central ignition was carried out, using COG-air mixtures with variable composition, at various initial temperatures and pressures. The main findings are summarized as follows:

1. Different explosion characteristics have been reported in a range from 298 K up to 423 K and from 0.5 bar(a) up to 1.0 bar(a)
2. The explosion pressures of COG-air mixtures attained their highest value at a concentration 1.0 within the studied concentration range of 0.25-2.25.
3. Most reliable ("worse case") explosion pressure is the pmax: 8.2 bar(a) for 298 K, 7.7 bar(a) for 323 K, 6.3 bar(a) for 353 K and 5.6 bar(a) for 423 K.
4. The explosion pressure, the maximum rate of pressure rise, the deflagration index are decreased, while the

normalized mass burning rate is increased with the increase of initial temperature.
5. The increase of initial temperature, the peak explosion pressure, combustion duration and flame development period all linearly decrease, while the maximum rate of pressure rise varies little.

This work would not have been possible without the financial support of Innovation for Efficiency and Environment - Growth, reg. no. LO1403 supported by National Programme for Sustainability and financed by the Ministry of Education, Youth and Sports.

References

1. M. Skřínská, J. Skřínský, P. Dolníček, P. Lukešová, , R. Přichystalová, Ch. Serafínov Mat. Sci. For., **811** (2015).
2. R. Razzaq, Ch. Li, S. Zhang China Fuel **113** (2013).
3. J. Skrinsky, J. Veres, V. Peer, P. Friedel, J. Travnickova, A. Dalecka Chem. Eng. Trans. **52** (2016)
4. Z. Zhang, B. Lin, G. Li, Q. Ye Safety Science **55** (2013)
5. EN 15967, 2012. Determination of maximum explosion pressure and the maximum rate of pressure rise of gases and vapours. European Standard, Beuth Verlag, Berlin Wien Zurich.
6. Dahoe A. E., Zevenbergen J. F., Verheijen P. J. T., Lemkowitz S.M., Scarlett B. J. Loss Prev. Process **9** (1996)

Model and Simulation of Permanent Magnets Synchronous Machine (PMSM) of the Electric Power Supply System (EPS), in Accordance with the Concept of a More Electric Aircraft (MEA)

Lucjan Setlak[1*], *Rafał* Kowalik[1]

[1]Department of Avionics and Control Systems, Polish Air Force Academy, ul. Dywizjonu 303 nr 35, 08-521 Deblin

Abstract. Based on the mathematical model of synchronous electric machine, basing on permanent magnets, presented in this paper, the key importance of alternator AC power sources in the form of generator (for conventional aircraft) and in the form of integrated unit starter/AC synchronous generator S/G AC (with respect to advanced aircraft concept in the field of more/all electric power MEA/AEA) was highlighted. In addition, through the analysis and selected simulations of the power supply system of a modern aircrafts, sources of onboard electrical energy (synchronous generator, integrated unit starter/AC generator) were located in board autonomic power system ASE (EPS, PES). Key components of this system are the electro-energetic power system EPS and the energo-electronic power system PES. Additionally, the analysis and exemplary simulations of key electricity sources based on mathematical models have contributed to highlighting the main practical applications in line with the trend of a more electric aircraft.

1 Introduction

Currently, gradually increasing dynamics of changes in the board power supply in modern aviation can be seen, both in the field of civil aircrafts (Airbus, Boeing) for airplanes (A-350XWB, B-787), as well as for military aviation (Lockheed Martin) in terms of aircrafts JSF (Joint Strike Fighter) F-35 and F-22 Raptor. These changes will have a crucial impact on rapidly growing trend in the more electric aircraft. They include in particular the dominant role of advanced systems in the production/processing of AC (Alternating Current) voltage and their basic components (rectifiers, converters, etc.), as well as the EPS (Electric Power System) in the field of DC (synchronous motor or a generator, a DC battery as a source of emergency, whichstores energy, fuel cells, etc.). Referringto the selected issues in terms of power electronics, power electronics associated with advanced power systems PES (Power Electronics Systems) it should be noted that the main components of the system PES, i.e. energy-electronic converters and their key elements, which are multi-pulse rectifiers are classified as so-called on-board autonomous electro-energetic systems ASE.
One of the most significant ASE systems, in addition to electric power systems EPS is a powersystemin the range of PES. Making a preliminary analysis of the dominant role of the on-board power supply voltage AC it should be noted that the power network is based on a generator of high frequency, namely, a Constant Frequency (CF) 400 Hz (in the case "classical" aircrafts) and Variable Frequency (VF) 390-780 Hz (in the case of airplanes "More Electric"). The above analysis includes also a combinationstarter/AC synchronous generator AS/G VF (Alternating Starter/Generator VF), an example of which together with key components of ASE system is ilustrated in the figure below (Fig. 1) [1], [2].

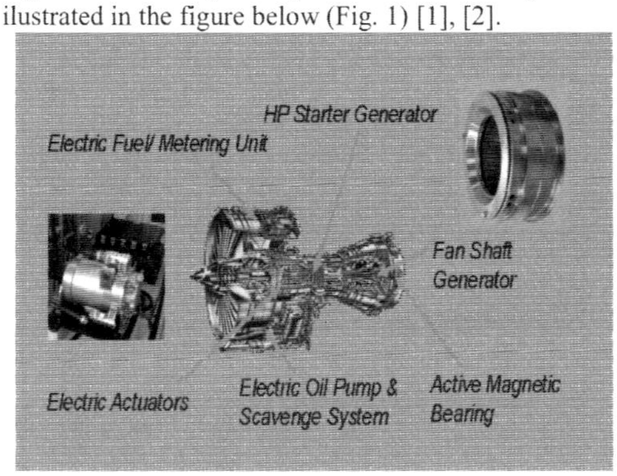

Fig. 1. Example of synchronous 3-phase AS/G VF in the context of MEE (More Electric Engine) [3]

On the modern airplanes there are primarily used tree-phase synchronous machine (eg. generators) as sources for generating electricity of multiphase alternating current of high power. It should be noted that this kind of electrical machinery PMSM works best during generator operation, that is, in terms of power. This does not mean that they can not operate within the motor, or in terms of starting the engine. However, in this operating range of their properties their operation is much worse. Typically, they are used in electric drive systems, which are

* Corresponding author: l.setlak@wsosp.pl, r.kowalik@wsosp.pl

https://doi.org/10.1051/itmconf/20181603004

required to maintain strict speed of specific systems, devices, etc. when the load changes in a wide range. The further part of the article presents the analysis along with the trend of more electric aircraft based on its main component, which is a 3-phase synchronous generator, serving as the primary source of electricity generation on board of modern aircraft.

2 Analysis and mathematical model of EPS of a More Electric Aircraft

Sample analysis was presented on the basis of a key component of EPS system, which is a synchronous generator PMSM with permanent magnets, which is a primary source of power for modern aircrafts complying with the trend of the more electric aircraft MEA, both civilian (Airbus, Boeing), as well as the military (Lockheed Martin) [4], [5]. Power supply of the board network, is done partly through its primary sources (generator, integrated unit starter/generator). The process of starting the engine is realized by a driving turbine and consequently mechanical energy is produced supplied by means of suitable systems to the generator. The task of generator is processing mechanical rotational energy of engine into electrical energy of 3-phase AC voltage, which is then fed to the head rail of autonomous power system ASE in terms of EPS system. It should also be noted that in the case of advanced aircraft in line with the concept of more/all electric aircraft (MEA/AEA), the boot process of electrical unit is realized by means of electrical energy, which is an important advantage of this type of solution.

Taking into account the high complexity of on-board power system EPS and its large load, the most suitable device (a source of electricity production) in the exercise of the function is a synchronous motor PMSM, having a good airworthiness in particular areas of work, autonomous work and robust design. Furthermore, the use of such a generator in aviation allows to produce not only 3-phase AC voltage, but also to obtain from it at least twice the electric power, comparing it with a DC generator in terms similar parameters, which are the size and weight. This property decided primarily about the application of this type of generator as a source of electricity generation on modern aircrafts because of the alternating voltage waveform generated by the 3-phase air generators. Sometimes colloquially they are referred to as alternators, but the term refers primarily to the sources of electricity generation in the automotive industry. The construction and method of cooperation of synchronous generator PMSM with on-board electrical system is presented later in this article. The following figures illustrate a simplified internal structure of the 3-phase generator PMSM (Fig. 2) and the general scheme of operation of the rectifier system (Fig. 3) [6].

In the 3-phase AC generator in terms of its architecture, in analogy to DC generator, you can extract the wound stationary stator, inside which a rotor rotates. Stator windings placed in slots are the same three phase coils, spatially shifted by an angle $\frac{2}{3}\pi$ rad. This kind of solution for the spatial distribution of the winding was obtained by placing the sides of each coil of one phase winding in the opposite the slots.

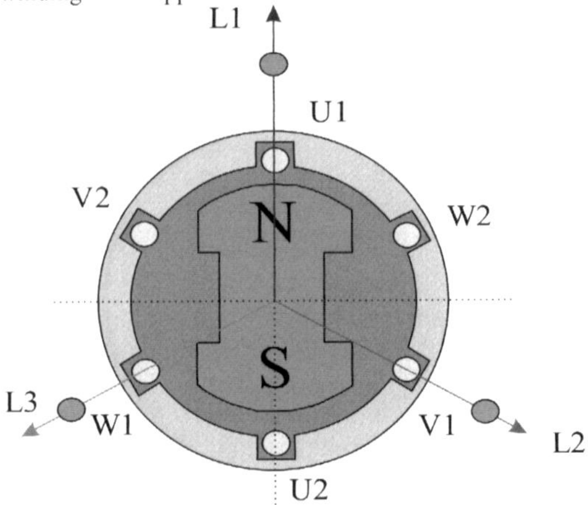

Fig. 2. The general structure of the synchronous generator PMSM

The source of the static magnetic field in a three-phase AC generator, as in the single-phase generator, is the rotor rotating at a constant angular speed speed ω within the three phases of wound stationary stator. In turn, the rotating constant magnetic field of the rotor during its movement induces in the windings of the stator phase voltages, which are called symmetric because they have the same amplitude and frequency, and differ only in phase-angle specified by spatial distribution of the phase windings on the stator of the generator.

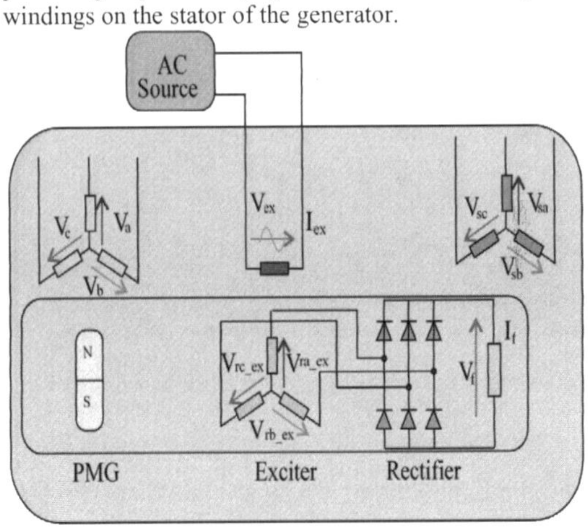

Fig. 3. The general functional diagram of the rectifier system used board of modern aircrafts

It should be noted that in terms of starter (motor) work, suitable torque must be made, which differs significantly from the load torque for the process to start the main engine of the generator. Therefore, the stator winding excitation is powered by AC, which via a transformer causes induction of the electromagnetic field inside the exciter, so it is possible to determine whether the rotor is rotating (working) whether it is permanent (idle).

Produced by this process, the electromagnetic field induces an AC voltage in exciter stator, which in turn by means of a rectifier is supplied to the synchronous generator rotor PMSM. Changing the frequency of AC voltage causes that the exciter is powered from the control of operation of the generator. In addition, the power output of the generator for the AC produces a torque which is delivered to the main rotor and the mechanical output power is obtained by rotating rotor. The Fig. 3 shows the wiring diagram of the synchronous machine PMSM on the field of motoring. It can be seen that the outputs of the synchronous machine are not connected, and the voltage of generator excitation is provided by an external source.

The equation describing the electrical phenomena in a synchronous generator PMSM in a *d-q* in the case of three-phase power system, with the included transformation of the position of the rotor *d-q* was expressed by the following mathematical relationship (designation used in the mathematical analysis is explained in Table 1).

The voltage in the position of the *d* [7], [8], [9]

$$V_d = R_s I_d + \frac{d\psi_d}{dt} - \omega\psi_q \tag{1}$$

The voltage in the position of the *q*

$$V_q = R_s I_q + \frac{d\psi_q}{dt} - \omega\psi_d \tag{2}$$

The voltage dependent on the frequency change

$$V_f = R_f I_f + \frac{d\psi_f}{dt} \tag{3}$$

Where the stream ψ produced in the synchronous generator in the position *d* is defined as

$$\psi_d = L_d I_d + M_{sf} I_f \tag{4}$$

For position to the *q*

$$\psi_q = L_q I_q \tag{5}$$

and

$$\psi_f = L_f I_f + \frac{3}{2} M_{sf} I_d \tag{6}$$

A further mathematical analysis of the electric model of PMSM generator includes saturation resulting in the system. Therefore, depending on the inductance value, the saturation rate K_{sat} can be written as

$$\begin{aligned} L_d &= K_{sat} L_{md} + L_{cd} \\ L_q &= K_{sat} L_{mq} + L_{cq} \\ L_f &= K_{sat} L_{mf} + L_{cf} \\ M_{sf} &= K_{sat} M \end{aligned} \tag{7}$$

where L, L_m and L_c denote the inductance, magnetic inductance and inductance losses of PMSM generator respectively.

It should also be noted that the ratio of saturation occurring in a PMSM generator is determined by determining the balance of amplitude magnetizing current, defined as

$$I_m = \sqrt{\left(I_f + \left(\frac{L_{md}}{M}\right) I_d\right)^2 + \left(\left(\frac{L_{mq}}{M}\right) I_q\right)^2} \tag{8}$$

where I_m is equivalent of the excitation current, obtained on a generator stator in a *d-q* and on the other side of excitation circuit.

The movement of the motor (turbine) is described by the equation

$$J\frac{d\omega}{dt} + f\omega = p(T_e + T_L) \tag{9}$$

where ω is a rotor speed of the generator; p is the number of poles in the PMSG; J moment of inertia; T_e the electromagnetic torque of the synchronous machine rotor designated in a *d-q*.

This value can be determined from the equation

$$T_e = \frac{3}{2} p\left(M_{sf} I_f I_q + (L_d - L_q) I_d I_q\right) \tag{10}$$

As it regards the moment T_L, it may be determined depending on the rotational speed. Moreover, the technical solutions of synchronous generator moment variation T_L is small in the time interval, therefore it can be assumed that a negligible effect on the dynamics of mechanical processes occurring in the generator it can be in further considerations omitted. Therefore, non-linear equation of state of synchronous generator in a *d-q* may be defined as

$$D = L_d L_f - \frac{3}{2} {M_{sf}}^2 \tag{11}$$

Complete (full) mathematical models, defining the status of work in the PMSM generator can be determined by performing the transformation of equations associated with the rotor (3) - (6). All possible differential equations describing the electrical phenomena in the electrical system, which is the generator are as follows [10], [11], [12]

 https://doi.org/10.1051/itmconf/20181603004

$$\begin{bmatrix} \frac{dI_d}{dt} \\ \frac{dI_q}{dt} \\ \frac{dI_f}{dt} \\ \frac{d\omega}{dt} \\ \frac{dT_L}{dt} \\ \frac{d\theta}{dt} \end{bmatrix} = \begin{bmatrix} -\frac{R_s L_f}{D} & \frac{L_q L_f \omega}{D} & \frac{R_f M_{sf}}{D} & 0 & 0 & 0 \\ -\frac{L_d}{L_q}\omega & -\frac{R_s}{L_q} & -\frac{M_{sf}}{L_q}\omega & 0 & 0 & 0 \\ \frac{3R_s M_{sf}}{2D} & -\frac{3L_q M_{sf}}{2D}\omega & -\frac{R_f L_d}{D} & 0 & 0 & 0 \\ \frac{3p^2}{2J}(L_d - L_q)I_q & \frac{3p^2}{2J} M_{sf} I_f & 0 & -\frac{f}{J} & \frac{p}{J} & 0 \\ 0 & 0 & 0 & 0 & 0 & 0 \\ 0 & 0 & 0 & 1 & 0 & 0 \end{bmatrix} \cdot \begin{bmatrix} I_d \\ I_q \\ I_f \\ \omega \\ T_L \\ \theta \end{bmatrix} + \begin{bmatrix} \frac{L_f}{D} & 0 & -\frac{M_{sf}}{D} \\ 0 & 0 & \frac{1}{L_q} \\ -\frac{3M_{sf}}{2D} & 0 & \frac{L_d}{D} \\ 0 & 0 & 0 \\ 0 & 0 & 0 \\ 0 & 0 & 0 \end{bmatrix} \cdot \begin{bmatrix} V_d \\ V_q \\ V_f \end{bmatrix} \tag{12}$$

It should be noted that the excitation current of the rotor is difficult to measure, because all electrical components are mounted together and very difficult to access. However, the field current can be determined from the measurement of the current in the exciter. With this in mind, in a very easy way it can be concluded that there is a constant gain (profit) current K_I which will assess the two excitation currents (I_f, I_{ex}) and it can be written as follows

$$I_f = K_I I_{ex} \tag{13}$$

This mathematical relationship describes the peak value of the AC generator excitation and represents a DC of generator rotor. In the analysis of power systems ASE (EPS, PES) of modern aircrafts you must take into account the variability of the mechanical turbine torque and the variability of environmental conditions in which the dynamic processes of components included in the on-board electrical network are implemented, hence the mathematical equations presented in discrete form. Therefore, a full numerical model in the space of the generator is reduced to the five states instead of six.

These matrices are as follows [13], [14], [15]

$$\begin{bmatrix} I_{\frac{d}{k}} \\ I_{\frac{q}{k}} \\ \omega_k \\ T_{\frac{L}{k}} \\ \theta_k \end{bmatrix} = \begin{bmatrix} 1 - T_s\frac{R_s}{L_d} & T_s L_q \frac{\omega_{k-1}}{L_d} & 0 & 0 & 0 \\ -T_s L_d \frac{\omega_{k-1}}{L_q} & 1 - T_s \frac{R_s}{L_q} & 3p^2 T_s M_{sf} K_i \frac{I_{\frac{ex}{k-1}}}{2J} & 0 & 0 \\ 3p^2 T_s (L_d - L_q) \frac{I_{\frac{q}{k-1}}}{2J} & -T_s M_{sf} K_i \frac{I_{\frac{ex}{k-1}}}{L_q} & 1 - \frac{T_s f}{J} & p\frac{T_s}{J} & 0 \\ 0 & 0 & 0 & 1 & 0 \\ 0 & 0 & T_s & 0 & 1 \end{bmatrix} \cdot \begin{bmatrix} I_{\frac{d}{k}-1} \\ I_{\frac{q}{k}-1} \\ \omega_{k-1} \\ T_{\frac{L}{k}-1} \\ \theta_{k-1} \end{bmatrix} + \begin{bmatrix} \frac{T_s}{L_{dt}} & 0 \\ 0 & \frac{T_s}{L_{qt}} \\ 0 & 0 \\ 0 & 0 \\ 0 & 0 \end{bmatrix} \cdot \begin{bmatrix} V_{\frac{d}{k-1}} \\ V_{\frac{q}{k-1}} \end{bmatrix} \tag{14}$$

and T_s is the sampling period.

Table 1. Symbols used in mathematical considerations

k/k-1	predicted quantity	L_c	leakage inductance
k/k	estimated optimal quantity	M	mutual stator/rotor
d, q	synchronous reference frame index	ψ	flux
s	stator quantities	K_{sat}	saturation coefficient
f	rotor quantities	K_i	current gain between I_{ex} and I_f
V	voltage	T_l	load torque
V_{dc}	DC voltage source	T_e	electromagnetic torque
V_{ZSSn}	zero sequence voltage	θ	electrical position
I	current	ω	electrical angular rotor speed [rad/s]
I_{ex}	exciter excitation current	J	inertia
I_f	main S/G excitation current	p	main S/G poles pairs
I_m	magnetising current	N	mechanical speed [rpm]
R	resistor	T_s	sampling period
L	self-inductance	L_m	magnetising inductance

3 Simulation of selected components of the EPS system according to the trend of MEA/AEA

At the beginning of the results of calculations we prepared a draft outline of the simulation model (Fig. 4). Presented earlier mathematical model was moved to the package of simulation program, which is the programming environment Matlab/Simulink. In the process of schema design of individual blocks the following assumptions were used presented in the Table 2.

 https://doi.org/10.1051/itmconf/20181603004

Table 2. Main Starter/Generator parameters 200 kVA, 230V, 300A. 400 Hz, three phases, Y connection, three pole pairs

Stator resistance	**Rotor resistance**
$R_s = 0.015V$	$R_r = 0.41V$
D axis stator inductance	mutual inductance
$L_d = 0.50$ H	$M_{sf} = 0.0037$ H
q axis stator inductance	nominal stator current
$L_q = 0.34$ H	$I_{sn} = 300A$

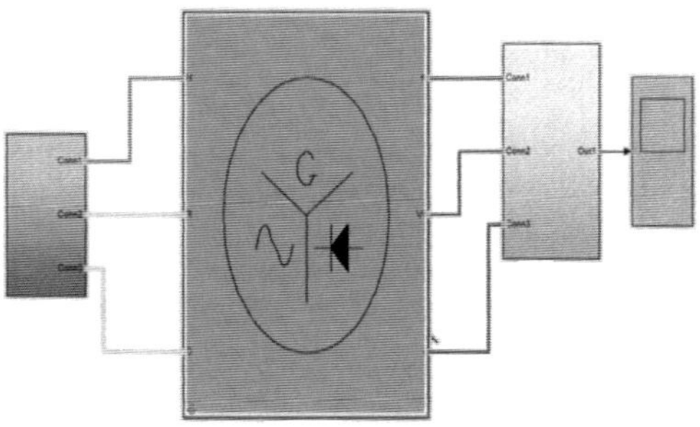

Fig. 4. The model of PMSM generator made in Matlab/Simulink

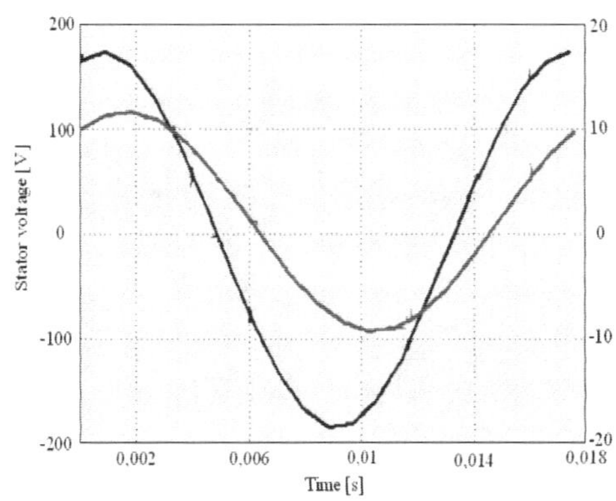

Fig. 5. The course of one cycle of the voltage and current AC on the stator

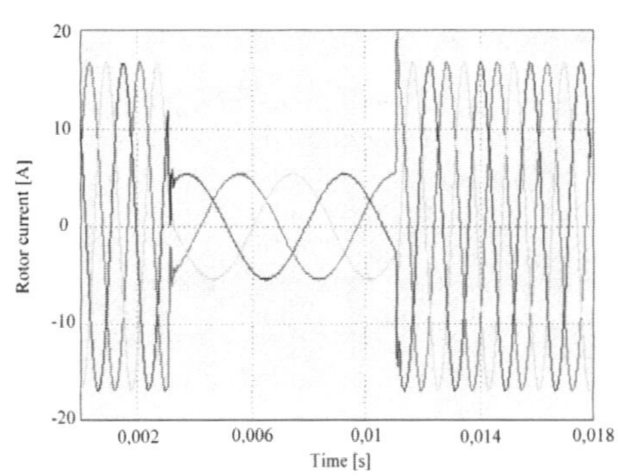

Fig. 6. Waveforms of 3-phase AC on the generator rotor

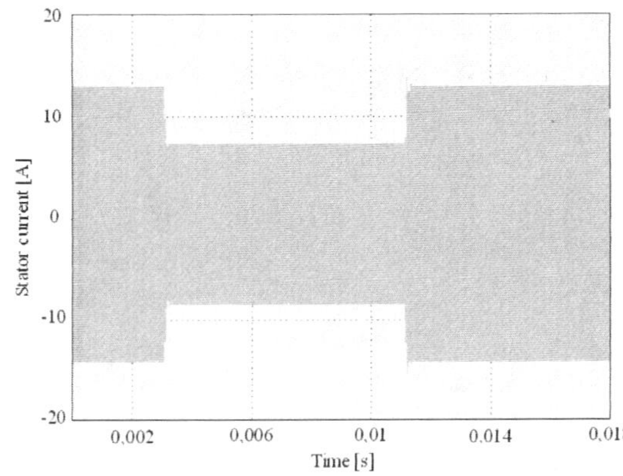

Fig. 7. Waveforms of 3-phase AC on the generator stator

It should be noted that the above voltage and current waveforms on the stator element of the generator (Fig. 5) supplied with a 3-phase voltage and current are distorted due to noise of electrical disturbances occurring in a generator and the temperature of the working environment of the device. There was also a significant variation of the amplitude signals and phase angle value for both voltages as well as AC. In turn, the waveform of 3-phase AC generator on the rotor (Fig. 6) are offset from each other by the same phase angle and have the same amplitude value and the period of repetition. Similarly, the situation is in the analysis of waveforms of three-phase AC on the generator stator (Fig. 7) [16], [17].

4 Conclusions

The article presents a study on the generator in terms of EPS system and mathematical modeling of electrical phenomena occurring in the production/processing of electricity in the airplane. The main study involved assessing the performance of the generator PMSM, supplied voltage and three-phase AC. The authors are deeply convinced that in the article solution to supply more electric aircraft will provide greater efficiency in comparison with "conventional" airplanes, equipped with a standard electrical network. The results of research, both on the basis of this mathematical model, as well as in the case of a simulation model showed that the assumed structural changes in the construction of the generator system were correct. First of all, they offer the possibility of obtaining suitably low value of AC voltage. Waveforms of harmonic currents in the induced voltage of the generator on the test rotor are defined by the current gain factor K_i. In addition, closer to the sine wave shape of the voltage can be obtained when the permanent magnets are arranged inside the rotor laminations.

Presented in the paper analysis of synchronous generator model with permanent magnets PMSM to increase the frequency of the generated voltage is to be an alternative mainly for generator sets, consisting of high-speed generators cooperating with emergency and low-efficient gearboxes and low-speed generators with a very large outer diameter of the stator. In every of these cases the new generator would get not worse efficiency, respectively high service life, lower weight and more compact unit structure (primarily in the case of a generator with a large number of poles).

References

1. Moir I., Seabridge A., Design and Development of Aircraft Systems, Second Edition, John Wiley & Sons, Ltd., 2013.
2. Setlak L., Kowalik R., Mathematical modeling and simulation of selected components on-board autonomous power supply system (ASE), in accordance with the concept of a more electric aircraft (MEA), 2017 18th International Scientific Conference on Electric Power Engineering, Kouty nad Desnou, Czech Republic, pp. 1-6, IEEE 2017.
3. Parker R., Horizon 2020 Industrial Technologies as an Engine of Growth, Rolls-Royce, Presentation 2012.
4. Emandi K., Ehsani M., Aircraft power systems: technology, state of the art, and future trends, Aerospace and Electronic Systems, IEEE, 2000.
5. Krause P.C., Wasynczuk O., Sudhoff S.D., Analysis of Electrical Machinery and Drive Systems, IEEE Press, John Wiley & Sons, Inc., 2002.
6. Setlak L., Kowalik R., Comparative Analysis and Simulation of selected Components of Modern On-board Autonomous Power Systems (ASE) of Modern Aircraft in line with the Concept of MEA/AEA, Lecture Notes in Engineering and Computer Science, Volume 1, 2016.
7. Yang Z., Qu J., Ma J., and Shi X., Modeling and Simulation of Power Distribution System in More Electric Aircraft, Hindawi Publishing Corporation Journal of Electrical and Computer Engineering Volume, Article ID 847624, 7 pages, 2015.
8. Abdel-Fadil R., Eid A., Abdel-Salam M., Electrical distribution power systems of modern civil aircrafts, 2nd International Conference on Energy Systems and Technologies, Cairo, Egypt 2013.
9. Singh B., Murthy S.S., and Gupta S., Analysis and design of STATCOM-based voltage regulator for self-excited induction generators, IEEE Trans. Energy Convers., vol. 19, no. 4, pp. 783–790, Dec, 2004.
10. Zhao X., Guererro J.M., Wu X., Review of Aircraft Electric Power Systems and Architectures, International Energy Conference (ENERGYCON), IEEE 2014.
11. Skvarenina T.L., Pekarek S., Wasynczuk O., and Krause P.C., Simulation of a More-Electric Aircraft Power System using an automated state model approach, Proceedings of the Intersociety Energy Conversion Engineering Conference, pp. 133-136, 1996.
12. Cheng R., Zhao W., Deng H., and Jiang X., Modeling and Optimization Control for Aircraft AC Generator Brushless Excitation System Based on Improved Adaptive PSO, The Open Automation and Control Systems Journal, vol. 7, 21-30, 2015.
13. Fernando W.U.N., Barnes M., and Marjanovic O., Direct drive permanent magnets generator fed AC-DC active rectification and control for more-electric aircraft engines, IET Electric Power Applications, vol. 5, no. 1, pp. 14-27, Jan. 2011.
14. Vadher Vinod Vashram, Mathematical Modelling of Aircraft Electrical Power Systems, Doctoral Thesis, Loughborough University of Technology, 1981.
15. Maalouf A., Idkhajine L., Le Ballois S., Monmasson E., Field programmable gate array-based sensorless control of a brushless synchronous starter generator for aircraft application, Published in IET Electric Power Applications, 2010.
16. Burrow S.G., Mellor P.H., Churn P., Sawata T., and Holme M., Sensorless operation of a permanent-magnet generator for aircraft, IEEE Trans. Industry Applications, vol. 44, no. 1, pp. 101-107, Jan./Feb. 2008.
17. Setlak L., Overview of Aircraft Technology solutions compatible with the concept of MEA, Technical Transactions. Electrical Engineering, No. 1-E/2015, p. 67-76, Cracow University of Technology 2015.

AUTHOR INDEX

EDP Sciences – Web of Conferences
17, Avenue du Hoggar
Parc d'Activité de Courtabouf
BP 112
F-91944 Les Ulis Cedex A
France

ISBN 978-1-5108-5576-2

CURRAN ASSOCIATES INC.
proceedings.com

XLIX International Symposium on Multiparticle Dynamics (ISMD 2019)

EPJ Web of Conferences Volume 235 (2020)

Santa Fe, New Mexico, USA
9 – 13 September 2019

edp sciences